Global Research on Urban
Planning and Development

世界城市规划与发展研究
战略规划 Ⅱ

马文军
Marisa Carmona 主编

中国建筑工业出版社

图书在版编目（CIP）数据

战略规划Ⅱ／马文军等主编．—北京：中国建筑工业出版社，2011.11
（世界城市规划与发展研究）
ISBN 978-7-112-13624-7

Ⅰ.①战…　Ⅱ.①马…　Ⅲ.①城市规划－研究　Ⅳ.① TU984

中国版本图书馆CIP数据核字（2011）第195694号

责任编辑：黄　翊　徐　冉　陆新之
责任设计：叶延春
责任校对：肖　剑　陈晶晶

世界城市规划与发展研究
战略规划 Ⅱ

马文军　Marisa Carmona　主编

*

中国建筑工业出版社出版、发行（北京西郊百万庄）
各地新华书店、建筑书店经销
北京嘉泰利德公司制版
北京方嘉彩色印刷有限责任公司印刷

*

开本：880×1230毫米　1/16　印张：14¼　字数：370千字
2012年9月第一版　2012年9月第一次印刷
定价：98.00元
ISBN 978-7-112-13624-7
　　（21369）

版权所有　翻印必究
如有印装质量问题，可寄本社退换
（邮政编码 100037）

序 一

世界已经进入城市的时代。城市的历史已有数千年，近现代的城市化进程也伴随工业化的发展而走过几百年时间，然而科技的进步已引领城市进入快速发展时期，可持续发展的要求、全球化及引发的竞争，对发展的渴求和掌握资源有限之间的矛盾，如此种种，迫使各国城市管理者思考城市应有的发展道路。

本书编者所带领的团队来自中国、荷兰、法国、日本、智利、阿根廷、巴西等十多个国家的高校及政府机构，在多年国际合作的城市研究基础上，采用比较研究的方法，分析了当今城市所处的世界经济、社会、政策环境以及城市发展研究的动态，对比了东西方城市的发展与规划实践。目前已研究完成的数十个城市分布于亚洲、欧洲、非洲、南美洲，覆盖了东西方国家、发达地区与发展中地区的城市。在深入地研究与分析后，作者按照相同的叙事框架，阐述了不同国家与地区的城市现象和背景的关系，探寻了它们面临全球化挑战时的应对战略，以及各自不同的发展路径选择及其成果。

这些成果涵盖了城市管理、政策、经济、环境、规划、开发等多方面的分析，为我们汇集和构建了城市规划与城市发展研究的案例库，既有利于我们了解与借鉴国内外城市发展的道路，总结各自的经验与教训，认识当今全球化背景下城市发展的路径，也有利于城市管理者与规划者选择适合自己城市发展阶段与发展环境的案例进行学习。

这其中的案例城市既包括了如法国、荷兰、日本等发达国家的城市，也有中国、巴西、南非、阿根廷等新兴经济体；既有巴黎、海牙、布宜诺斯艾利斯这些国家首都，也有里约热内卢、台北、毕尔巴鄂这样一些重要的区域性中心城市，以及福冈、罗萨里奥等中小城市。这些城市的发展历史跨越了席卷全球的 1930 年经济危机、2001 年拉美经济危机、石油危机、亚洲金融危机、美国次贷危机等时期，反映了不同战略实施的过程与周折，能够帮助我们在坚持持续稳定的战略下，即使面对纷繁变换的世界环境，也要从容应对各种环境因素的变化。

同时，随着中国经济与社会发展成就日益获得世界瞩目，有关国内城市发展的经验也同样得到世界各国与地区的城市的关注，值得我们的研究者适时地加以整理，使国内城市发展的一些成功经验，也能够为其他发展中国家的城市提供借鉴探索适合的发展道路，为人类社会的健康发展作出贡献。

编者制定了包括 300 个中外城市的宏大研究计划，覆盖更多国家与地区的城市样本，当能为城市研究发挥更大的参考价值。是以为序，以为鼓励！

苏东水

（苏东水，复旦大学首席教授，国务院学位委员会学科评议组成员，世界管理协会联盟 IFSAM 常务理事兼中国委员会主席，中国国民经济管理学会会长，上海管理教育学会会长）

序 二

2012年2月9日中国社会科学院发布《国际城市发展报告2012》指出：2011年中国的城镇人口比例达到51.27%，从统计学意义上，中国已成为"城市化"国家，2012年成为中国城市化的"元年"。这表明中国几千年农业文明的农民大国，进入城市社会为主的新发展阶段。

城市的健康发展影响到整个国家和社会的发展，以及人们的生活。尤其当今世界局势，一方面经济全球化、国际劳动力分工，世界贸易在增加；另一方面，自从2008年以来的全球性经济危机，正改变世界全球经济格局，对市场和资源的争夺越演越烈。这些都使得城市不得不面对更为复杂的竞争与挑战。

对于城市的发展来说，要想在这竞争中取得优势，制定并实施一个科学的城市发展战略必不可少。因此，应广泛地借鉴世界各地的经验，如城市管理体制，一些国家的战略规划已推动了立法、公共层面的调整，并设立了国家和地方委员会来开展战略性的规划等。

城市大规模开发项目（建设能源系统、高速公路、多节点的运输终端、集装箱码头、大型化和现代化的码头、机场、空运设施等方式来改善基础能源、交通和电信基础设施的服务等）被城市战略赋予了重要的角色，即刺激城市发展和再生，吸引外国资本和人力资源，也被视为城市生产力发展的重要保证。

了解国内外城市的经验与战略，既有利于我们学习借鉴，也能帮助我们的城市做到知己知彼，在日益激烈的城市竞争中赢得有利的地位。

然而，城市之间问题千差万别，经济发展程度不同，地理气候和文化习俗迥异，管理体制也相去甚远，发展历程与阶段亦有区别。这些都阻碍了我们进行相互间的直接借用，而案例研究方法具有不可替代的优势。

本书是在多年国际合作的基础上完成的。本书著者是一个国际团队，他们来自各案例城市，参与或了解相关城市发展与规划的背景及其编制过程。系列性的城市对比研究，有利于我们在了解案例城市的战略制定、实施及成效的同时，在对比中判明其特点和思考其背景的差异性，知其然，又知其所以然。

为帮助读者进行比较分析和研究，本书19个案例的体例都相同。从城市基本数据、城市概况、城市发展历史与规划历史、全球化及相应挑战、城市发展战略、城市大规模发展项目等方面展开讨论。通过相互间的比较，展现各案例城市发展战略的形成背景与实施过程。

本书展现了跨地域、国际化、本地化的特点，构建了世界城市发展与规划的比较研究案例库，兼具理论性与工具性的作用。在我国进入城市社会为主的新发展阶段，面临着新问题，借鉴世界各地的经验是十分必要的。其中绝大部分的案例是来自发展中国家的案例，十分难得。如此集中地介绍世界众多城市的经验，对扩大思野、提高理论素养一定有所裨益，值得一读。

陈秉钊

2012年5月7日于同济大学

（陈秉钊，同济大学建筑与城市规划学院教授，曾担任中国城市规划学会副理事长、教育部科技委员会学部委员、建设部城市规划专业指导委员会主任、全国城市规划专业评估委员会主任）

自　序

20世纪无疑是城市化的世纪，世纪之初只有13%的世界人口居住在城市，而世纪结束时有一半的世界人口加入这个阵营。21世纪这一过程将会延续，而有关全球城市化重要意义的意识将会更加提高。

或许更加值得注意的是全球视野内所谓"新兴国家"的出现，伴随而来的是全球生产系统的改变、世界贸易的增加、国际分工的改变、国家和市场的关系，还有大城市群和全球城市在世界经济中的核心作用。信息和技术革命带来的时间和空间的压缩，在城市和城镇体系的组织以及形成中的全球通讯、流动性、知识和创新中提供了极其重要的角色，这也是这一进程与19和20世纪基于工业发展和现代化运动的城市构成的区别。

最新的全球资本主义发展在世界各地的建筑、规划和城市设计实践中产生了相当大的影响。战略规划被看作是城市利用全球化和城市大规模开发项目的机遇，并作为贯彻基于全球策略的未来发展观的主要机制的最好途径。这使城市干预的规模产生了剧烈变革，促进了复杂城市管理系统的发展，提高了城市项目的竞争力，同时也导致了更大的社会和空间碎片产生，并且使环境更加脆弱。

同时，战略规划也可以帮助城市通过获取这些大规模的干预所产生的增加价值，为"公共增益"开拓新的前景；也产生一种普遍的看法，认为存在于国家与市场关系中的某些东西需要进行审核，以使这些收益的分配成为可能。此外，由于现代化运输和电信基础设施的发展，空间和时间的大幅度压缩，使城市功能的向上标量转移，尽管空间和时间的流动性和成本差异会随之而来。在更广泛的区域范围内，已出现工业制造活动的减少和所有活动的普遍无计划扩展。

盛气凌人的桥梁和机场、优雅的大道、惊人的建筑在很短的时间内出现在世界各地。对高度的追崇，形式的奇特和"震撼效果"的教化，重塑了世界各地城市的天际线。一位著名的荷兰建筑师设计的高楼从美国延伸到中国，再回到他设计的一幢墨西哥的100层高楼；在毕尔巴鄂，一位美国建筑师设计了引人注目的古根海姆博物馆，另一个类似的在巴拿马；在马来西亚一位阿根廷建筑师设计了最高双塔，另一个则在海牙的中心。全球化，似乎正在屋顶咆哮着。建筑、工程和项目管理已经在接替规划学科来负责城市的形式与自然的关系。通过对大型项目的干预，这种城市规划新途径已经出现，这可以更好地被定义为"大型项目相关的规划"。对一些人来说，这是极好的；但对另一些人来说，它带来的还有不可持续和社会分化。

在过去，作为对帕克、格迪斯、芒福德、柯布西耶、卢西奥·科斯塔、奥斯卡·尼迈耶等人构想的体现，城市愿景再次与带着虚构目标的大城市，如美丽城市、花园城市、法兰西风情城市和魅力城市的转型联系在一起。21世纪，国际贸易委员会掀起的技术革命使全球进入新的发展阶段，与此相关的大规模城市项目日益受到关注。有关城市愿景的探讨致力于普通城区向信息化城市、全球化城市、一流城市、创意城市、文化城市、智能城市、网络城市、知识城市、生态城市、友好城市等等理想城市类型转变的议题。无论选择什么样的愿景，提高区域竞争力和城市对全球化开放性优势利用率的基本需要始终存在。大城市集群在实现基于加速和加大定制产品生产规模，在灵活金融资本和先进的全球服务节点流量增加所带来的新的全球经济体系相关的规模经济压力下，产生了新的城市系统和城市结构。这种新的、扩散式的、四分五裂的多核系统通过快速机动性走廊和高效运输系统结合在一起，根据最新的愿景构想重组成为现代区域功能型网络，从根本上改变了源自工业、现代主义和凯恩斯主义的旧区系统以及城市结构。新功能出现的同时，其他功能逐渐被摒弃，老的城市被大面积地废弃，同时由于城市化区域通常成为新的全球性活动的举办地点，各类土地用途的兼容性也已经发生改变。这种分散的、凌乱的、有核的新系统被快速的流动通道和高效的交通系统限制在了一起，依据假设的观点，新系统正重组着现代的区域性实用网络。这从根本上改变了从工业化、现代主义者、凯恩斯理论的时代继承下来的旧有的城市系统和城市结构。新的城市功能出现的同时，其他的功能就会过时，大片的城市区域随之被抛弃。因新的全球活动在城市化区域内开展，各种各样的土地使用的适用性也已经发生改变。

自从20世纪80年代以来，弗里德曼和沃尔夫（Friedmann & Wolff，1982）、弗里德曼（Friedmann，1988）、哈卫（Harvey，1989）、萨森（Sassen，1991）和卡斯特尔（Castells，1995，2001）这些技术决定论者解释了如何通过信息革命来实现新时期的资本积累。在这其中，他们把跨国公司当成具有领导地位的角色，跨国公司改变了全球劳动力分配，重组了各种水平和尺度上的空间。先前的全球制造系统是基于进口替代措施、联合贸易保护的工业化、福特制（Fordism，一种使工人或生产方法标准化以提高生产效率的办法，始于福特汽车公司）和凯恩斯主义理论的混合经济。这种经济正被以自由市场、基于出口导向型的全球新自由主义、大宗消费产品和服务的生产为特征的新系统所取代。

专业的金融服务、电信、基础设施和工业服务的发展，已经刺激全球贸易、投资、金融、科技转让以及劳动力的流动。这些发展对作为地区成长核心的大城市十分重要，这些服务与资金的提供极大地强化了全球的经济流动。

按照科技决定论者的观点，相对20世纪60和70年代来说，知识创造、创新和生产率在空间上发挥着更大的作用。有人认为，当市场开放、规模扩大，大型城市聚集区就自然而然地成为金融调配和资本聚集的中心。萨森（Sassen，2000）认为"城市成为了战略节点，通过此节点可以规划和促进新经济"。卡斯特尔（Castells，1995，2001）提出了"信息化城市"，哈尔法尼（Halfani，1996）提出了"象征性经济"，他们认为在创造附加值方面建筑和规划的作用增强，而资本投入的作用则会降低，应该提高社会基础设施、教育、物流以及创新的地位。

新的全球金融地理重新强调了大的城市聚集与都市圈的作用，特别是大多数发展中国家的政府提倡全球化作为摆脱欠发达的最有前途发展方式（Toledo-Silva，1995）。他们预计在城市的现代化和经济增长之下，全球化将最终导致社会和经济发展传播到所有区域和社会族群。这就是说，随着区域和城市对国际资本开放它们产品和金融服务，并日益成为占支配地位的经济单位，国家的边界正迅速失去其重要性。这些社会和空间变化的速度和剧烈程度将会区别全球化的不同阶段，并使得分析全球化、战略规划、大规模城市项目之间的关系成为迫切的议题。

本系列著作是关于全球化背景下世界各城市发展、管理和战略规划的研究成果。它提供了信息和对城市化进程的分析、城市战略的历史性发展，以及数十个案例城市中作为新的规划方式出现的大规模城市项目。它探索了6种新的造城方式，其核心要素包括：

- 通过项目建设而实施城市规划；
- 土地的社会性功能发生改变，土地及房地产的增值为私人占有；
- 城市功能的私有化和服务的公共特征减弱；
- 从城市尺度的综合城市管制，向着在标准化的框架中制定特定规章的转变。

在书中分析的大多数城市中，处于统治地位超过50年的国家和市场之间的关系已经变化，同时也有规划、融资、设计、管理城市的途径。本系列著作的目的是为了显示大规模城市项目演变成为战略规划的主要部分，也是为了更好地分析同一时代全球化趋势和发展战略的关系。

然而，大规模城市项目带来了一系列重大问题。首当其冲的是：这些项目能否在满足短期社会需求的同时也满足长期的社会需求？作为全市性城市诊断机制的大规模城市项目，能否也被用作协商工具，用以建设一个更加民主的城市环境？大规模城市项目如何应对由它们自身带来的社会不平等和环境损耗的外部负面效应？决策者们应当如何在城市信息不足、贫困社区不愿参与以及财务负担巨大这样一些背景下处理战略规划和大规模城市项目的问题？能否通过规划的大规模城市项目形成新的城市形态，提供应对气候变化和能源消耗问题的可持续解决方案？

当前，在富人们居住的富裕、舒适和安逸的环境之外，还存在着提升城市生产力所带来的环境问题、城市扶贫问题，以及减少成倍增长的暴力和不安全问题，城市内部的社会福利和发展机会鸿沟不断加深。伴随着贫困现象的城市化，特别是在发展中国家，反映出城市不仅是国家经济增长的引擎，同时也是穷人贫困和物质匮乏的场所。1995年的一份报告曾经预测到2000年全球将有10亿人每天生活费不足1美元，2007年的联合国人居中心全球人类住区报告中预测在2010年该数字将达到20亿人。

与新形式的战略规划相关联，全球化的一个特点是城市空间分裂、社会分化，以及地区之间、国家之间、城市和社会团体之间的不平等现象迅速发生（伯吉斯，2004，2005；布伦纳，2004；格雷厄姆和马文，2001），在发达国家和发展中国家之间和这些国家内城市的多样化现实情况正在形成。因此，这本书也探讨了地方和区域发展之间的关系、经济增长和社会发展之间的关系，以及贫困现象和现行政策之间的关系。

这项研究涵盖了不同经济和社会发展水平的国家的城市。本书所包含的19个城市来自于在欧洲、亚洲、南美洲、中美洲和非洲的15个国家，这其中包括位于世界体系的核心、半边缘和周边的城市，根据他们在新的全球劳动空间分工体系中的不同结构位置来解释这些城市彼此间的差异是最佳的解释方式。毕尔巴鄂、巴黎和海牙这一系列的欧洲城市，其中大部分经过了一个人口收缩和经济衰退的过程，这些过程反过来成为它们城市再生政策的基础。墨西哥城在一系列的北美城市中，作为世界最大的特大城市之一，尽管受到了权力分散的影响，但仍然取得了持续的增长。书中涉及的南美城市在人口和经济发展方面呈现出不同的情况。一些城市人口减少，但经济增长；一些城市人口增长，但经济却持续衰减；还有一些城市则是在人口与经济方面都出现锐减。这些城市包括波哥大、加拉加斯、基多、圣保罗、里约热内卢、布宜诺斯艾利斯、罗萨里奥、圣地亚哥。被研究的亚洲城市大多数呈现出蓬勃发展的态势，这些城市的人口和经济出现了快速增长，这些城市包括福冈、上海、新加坡、雅加达和台北。非洲城市则包括蒙得维的亚和卢萨卡。

仔细观察就能发现这些国家和城市之间巨大的经济差异。荷兰、法国、日本和新加坡等国家的人均国内生产总值大约是40000美元，阿根廷和智利分别是6000和9000美元，南非为5700美元，中国则为2500美元，而赞比亚的人均国内生产总值不到1000美元。各个国家的城市在贫困率、收入分配、生活质量以及文化、政治和体制方面也呈现出差异。

本书包含的城市之间有着显著的差异：它们中有些是大城市，而有些则是在大城市地区的小城市；有些是一线城市，而其他则是二线城市；有些是港口城市，而其他的则位于腹地；有些是正在衰退的老城市，而另一些城市正在迅速扩大；一些城市正处于逆都市化的进程中，有的则向外扩张，吞没现有定居点。一些城市人口规模小——如海牙、鹿特丹和瓦尔帕莱索都不到100万居民；而另一些特大城市，如东京、圣保罗、墨西哥城、里约热内卢、北京和布宜诺斯艾利斯的人口、空间范围和经济优势都仍在不断扩大。历史因素是城市在当地政治、决策和地方治理方面存在巨大差异的原因。

不同的国家和不同的城市之间的社会、政治问责制和城市民主存在很大差异。本书中的城市形态和结构涉及不同的历史进程和不同的原始积累方式，以及地方精英把从经济活动中获得的盈余投资（或不投资）到城市的不同方式。许多欧洲城市原始积累的案例与殖民主义有关，在纪念性建筑、社会服务和基础设施建设对城市结构的影响中表现出来。许多拉美城市也由殖民利益塑造而成。然而200年的独立连同地方势力一起以独特的方式塑造着城市。在非洲国家，殖民主义的影响仍然相当持久——城市当前的规划和安排方式就能清楚地反映这一事实。然而，总体规划的持续影响、政府对城市快速扩张和人口流动的反应，以及城市发展计划中的双边和多边援助的重要性，都有助于城市在处理它们的过去和既有的建成环境，以及顺应全球化中采取不同的措施。南非城市与其他大陆城市不同，是因为它们独特的城市发展历史、以前城市战略的特殊性以及国家权力在城市生活各个领域起的决定性作用。资本积累的种族隔离模型对全面发展规划和民众参与规划过程的空间影响持续数十年之久。

　　本书的案例部分由19个城市的独立章节组成。每个城市使用相同的格式进行系统性分析：首先是在介绍城市的基本数据，接着是城市简介，概述了当前发展的主要特点；城市历史；规划背景；全球化对具体城市的影响分析；对近期或目前正在实施的战略规划和城市主要工程的作用进行分析；再接下去是对最重要的大规模城市项目的深入研究，分析涉及其中的利益团体和个人，最后得出结论。

　　尽管这19个城市在地理、环境、社会、经济、科技和文化等方面存在着差异，但它们也存在很多共同点，需要加以研究以得出一些结论。这些共同点包括因全球技术和文化发展推动的相似的城市化进程。最后，确定了三个一般性问题，即大都市区（城市区域）甚至更大的集合城市的形成；提升城市竞争力的方法，包括技术和社会基础设施的改善；最后，完成从工业社会向服务社会的转型，包括城市形态、城市功能和管理工具的改变。

目　　录

总论 ·· 1

第1章　全球化时期的世界城市规划 ··· 3
第2章　经济危机冲击下的城市发展——营销导向、"精致城市"及其他 ······················· 14

案例 ··· 23

毕尔巴鄂（西班牙） ··· 28
海牙（荷兰） ·· 53
基多（厄瓜多尔） ·· 71
里约热内卢（巴西） ··· 87
罗萨里奥（阿根廷） ··· 106
墨西哥城（墨西哥） ··· 126
上海（中国） ·· 144
圣地亚哥（智利） ·· 163
新加坡（新加坡） ·· 183
雅加达（印度尼西亚） ·· 200

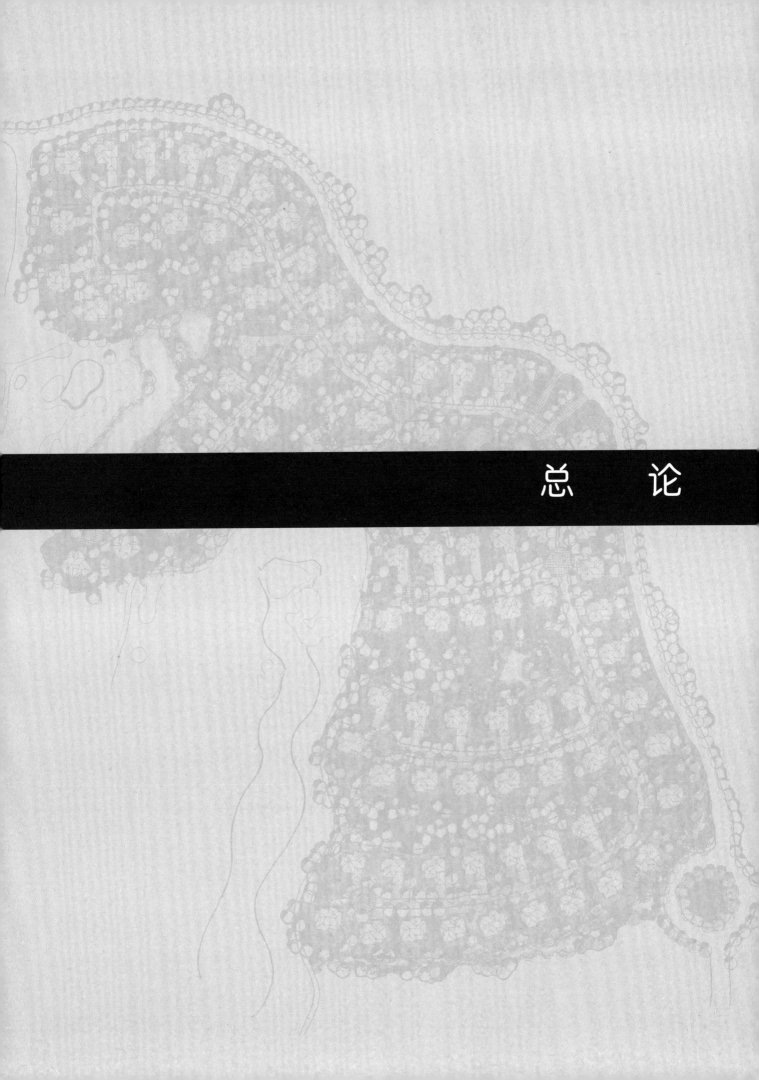

总　论

第1章 全球化时期的世界城市规划

马文军　玛丽莎·卡莫娜（Marisa Carmona）

1.1 概述

毋庸置疑，20世纪是城市化的世纪。20世纪初，世界上只有13%的人口生活在城市，100年后城市人口大约占到全球人口的一半。在21世纪，我们将看到城市化的进程还会延续，城市化产生的效益会相应提高，人们还将意识到全球城市化的重要性。

同样重要的，无论是位于全球系统的核心、半边缘还是边缘的新老城市，其规划方式已发生变化。由于全球化进程的加快和全球新自由主义发展战略主导地位的巩固，引起了经济、社会、空间和环境的调整，进而带来了各国城市规划理论和实践的重大变化。这些变化可以归结为城市规划基本模式的转变，可以定性为从总体规划到战略规划的转变。

本书旨在研究和分析这种向战略规划的转变和其主导的"操作方法"，即通过大规模项目或战略性项目的实施而实现规划。为此，本书第1~2册运用案例研究的方法，在全球范围内选择了20多个分别位于全球网络中的核心、半边缘和边缘地区的城市作为案例进行剖析。

当然，我们必须清醒地认识到从总体规划向战略规划的转变仍是一个未完成的过程，有些国家和城市走得更远和更快些。世界上不同地区的城市或同一地区的不同城市间继续存在着城市政策的差异，经济、社会、政治和文化的不同更是不足为奇。但在相同的国际背景下，发展的主流方向是一致的，即向战略规划转变。事实上，现在有一些人认为世界各地的政策已经达到了历史上前所未有的统一程度。然而，若要理解从总体规划向战略规划的转变和"通过项目建设实现规划"的适用性则需要进一步探讨近50年来城市及空间规划的理论和实践与主要发展战略间的关系。

另一方面，案例研究方法在法律、医学和管理领域的运用都十分普遍，但是在城市规划方面却刚刚起步。从案例研究方法的应用中，我们可以发现，通过大量资料的收集，分析复杂的现象，案例研究方法能够使我们获得优于其他研究方法的经验知识和数据。本书结合我们的研究过程，对城市规划案例概论、方案设计、资料收集、证据分析、成果表达和后评价等各个环节进行了研究与总结，系统地阐述了城市规划案例研究的方法。

1.2 关于城市的"第二次现代化"发展概念框架

在城市的"第一次现代化"（工业化初期）发展中，大城市充满了混乱与疾病。相反地，第二次现代化时期，城市发展被视为对国家经济发展起到了重要作用，而城市大规模项目成为城市转化过程中的重要动力，是"一个公共或私营的干预活动，因为其规模对城区、城市乃至区域整体上带来影响与冲击作用"（Iracheta，2000），是指"为了适应与推动城市经济的发展和城市功能结构的优化，为了满足人们不断提高的生活、工作、交通和文化娱乐的需求，以及出于政治、社会经济发展战略以及防止战争与自然灾害等因素的考虑，由政府或开发机构通过财政投资、银行贷款、发行债券等社会融资形式和通过开发机构自筹资金组织兴建的大型城市建设工程项目"（马文军，2000）。作为独立而统一的项目，城市大规模项目需要经过规划、设计，并按照确定的计划分期实施，有不同的利益方和明确的目的与目标，有一个负责的管理机构，对于成本、收益及社会、环境影响有预先的评估。

为了确立城市大规模项目的目标，首先需要为城市的两次现代化过程进行概念界定。

如人们所知，在新时期的全球化发展中，城市的第二次现代化不再被看作社会疾病的原因而被视为发展的动力（Beck，2000）。城市使农产品的价值得到提升，为区域性市场提供服务，并且吸引制造业与服务业领域的投资。高水平的城市化意味着较高的人均GDP、较高的妇女参与率、较高的教育和技能水平。在第二次现代化中，城市规划政策的制定贯穿着新自由主义①的思想，以提高城市生产力和市场化运作作为前提，包括了私有化、减少管制、权力分散和提高公共机构管理能力、增进各种社团的参与度（World Bank，1990）。它已经导致了政治上的分权、管制减少以及民主化进程，目的是推动新的管理、更多的市民参与以及全社会"达成共识"。

由于城市发展对国家的发展起到重要作用，因此其对城市生产能力的制约将直接影响国家经济的增长，这些制约包括：

（1）技术基础设施覆盖率、管理及维护上的不足，使得城市的生产成本提高、劳动生产率降低和竞争力下降。

（2）国家对于土地及住房市场的管制过度或者不够对市场供求方都有影响，既阻碍了对基础设施的投资规模的扩大（供应方），也影响基础设施用户数的增加，并因此造成政府不得不提供补贴。

（3）住房与服务方面的低效与不足会降低劳动生产率，而过度管制非正规经济将减少就业率及市民创造收入的机会，造成城市管理能力和技能的不充分、中央政府与地方政府间不合适的权力分配、脆弱的地方税收能力、不合适的补贴与不充分的城市开发融资服务。而确保城市生产率的根本提高所需要提供的大规模主干基础设施、服务管理、扩大覆盖率及改善维护问题的投资却无法得到满足。

（4）城市竞争，即城市对资源、市场、机遇的争夺。在全球化趋势的影响下，随着贸易自由化和信息技术的发展，以及各国中央以及上级权力的下放，城市竞争成为提升资源禀赋的重要途径。在计划经济时期，城市命运与中央各部委、省级各部门的重点项目与资金紧密联系，城市之间进行着简单的横向竞争，所谓跑"部""钱"进即为此种现象的生动描述。在转型期乃至将来的市场经济时期，城市不仅多了更多触角，并且可以伸得更长，在城市发展所依赖的土地、劳动力和资本总量有限的前提下，城市竞争的内核和形式发生了剧烈的变化。

要消除所有这些制约，提升城市整体竞争力，需要政府用市场导向的战略，制定恰当的鼓励措施，通过市场化的价格机制消除市场瓶颈，通过私营机构提供广泛的城市公共产品及服务。土地、住房、金融、基础设施、服务及劳动力市场的改革和自由化则成为提高城市生产率及运营效率必不可少的手段。新的空间与城市政策上的变化促进了城市大规模项目的出现；政府开放经济、吸引投资及改变城市面貌的政治意愿则加速了城市大规模项目的应用。

1.3 第二次现代化时期城市大规模项目的目标

城市大规模项目是新自由主义思想下城市发展政策的组成部分，即使城市由制造业为主向服务业转变，

---

① 新自由主义强调政府的作用是维护市场机制合理运作的秩序。

同时也改变了城市的空间格局，使中心商务区（CBD）的重要性越来越高，旧城区的复兴与活化[1]成为发展中越来越重要的部分。随着国际性竞争的加剧，公共投资已经被视为激活发展并阻止旧区持续衰落与恶化的关键[2]。

在这样的变化环境中，政策制订者必须仔细研究城市的历史、文化及社会传统，研究城市的起源、设施、产业，以期正确地把握自身拥有的特色和优势。城市大规模项目的目标就是借助自由市场带来的机会，以最小的投资，对城市结构产生最优化的影响。

同时，大规模项目试图将城市放在全球的背景下加以定位，通过新的科技、基础设施及服务，吸引国内和国际的投资者。对于正在衰落的城区，大规模项目可以形成新的发展动力，并形成多层次的乘数效应，推动城市竞争力的提升。Oriol Bohigas 在论述巴塞罗那 Vell 港的经验时候说："它（大规模项目）通过一系列战略性的城市运作，一旦完成后将推动城市在结构与功能上实现新的现代化。"一个始于小尺度的发展最终实现了大都市的框架（Garay，2002）。

不仅旧城改造逐步深入，城市边缘也开始得到重视。通过整合并借助土地利用、交通可达性及交通方式变化的综合研究，过去空置的土地也找到了新的用途。显然，环境改善目标与促进中心区和边缘地区经济与功能的均衡发展获得了统一。

大规模项目与环境保护的内容同时整合于城市更新计划中，使可持续发展目标成为可能。

大规模项目也可以与公共及私人机构投资的住宅项目共同进行。为了改善旧城的居住功能，提高流动性和生活品质，大规模项目可以创造更安全、多样化及对外资充满吸引力的城市环境。

1.4 城市大规模项目的特点与类型

城市大规模公共开发项目的空间规模、时间跨度、辐射区域范围都远远超过一般建设项目。其项目的特点反映在项目的决策、规划、实施及运营的各个阶段，并在一定程度上决定了项目策划及实施过程中的特征。

城市大规模项目建设的特点包括：

（1）建设规模及投资巨大、配合的环节众多

城市大规模公共开发项目的建设规模巨大，在项目的投资、组织管理、占地面积、工程量、所需机械设备、技术及劳务人员等方面，较一般项目的投入大得多。

（2）技术及社会风险大[3]

城市大规模开发项目由于工程规模巨大，影响因素难以全面识别，技术问题非常复杂，由此带来的技术风险和非技术风险都较一般项目大。在国外，很多项目甚至成为不同政党派别政治角力的重要筹码，如加拉加斯的琼娜伦特城市广场项目。

（3）项目的生命周期长，在城市经济和社会发展中占重要的战略地位

城市大规模开发项目是从社会进步和经济发展的宏观战略角度考虑立项建设的，由此需要进行客观、全面而可靠的可行性研究，其建设也必然是一个较长期的过程；同时，着眼于全球化背景下城市长远的经济、社会利益，大规模开发项目建成后的运营、提供的社会服务也是一个长期的阶段。

（4）对自然环境和社会环境有重大影响

城市生态环境和城市大规模项目相互构成对方的风险因素。一方面，由于大规模项目规模巨大，涉及

---

[1] 王训国，马骥. 都市再活化——上海虹桥地区功能完善与拓展研究 [J]. 规划师，2005（6）.
[2] 青岛市政府东迁10余年，吸引各方资金加入到城市新区的建设中，把原来东部农村变成了新城区、新的城市中心和中央商务区，形成了以香港路为中心的行政中心、金融中心、经济中心、旅游景点，使青岛新市政府、五四广场成为新青岛城市的标志，并一举改变了百年来青岛城市南北狭长的带状结构，使青岛城市顺利向东拓展，优化了城区几何结构。
[3] 参见：福州青州造纸厂15万t本色纸浆项目建设失败案 [N]. 科技日报，1998-06-20.

区域多，在项目的建设过程中有可能严重影响建设地区自然、生态、卫生等环境的质量状况；建成后，对城市现有物质环境和人文环境的影响更是长远的。另一方面，城市生态环境的改变也会影响项目正常的运营，而人文环境的变化更是可能危及项目存在的根本条件。如为了改善交通条件和生活条件而建造大量的道路和公交设施，一旦大量人流疏散到环境质量大大提高且交通仍然便利的郊区，不免会带来中心区人口的流失，最后使商业设施与居民一起外迁，造成中心区的衰落，同时当初建造的那些道路交通设施的用户必然会减少，效益也会下降。近些年上海大规模的城市旧城改造，使得中心区人口大量外迁流失，黄浦、卢湾、静安、虹口这几个城市中心区人口有3%～15%的减少（表1-1），原来中心区的商业服务设施已经面临着客流的大量减少，销售增长趋缓（表1-2），有些甚至出现了持续的下滑（如黄浦区）。这种国外一些发达国家城市出现过的"空心化"（Empty Donut）[1]的现象，如美国纽约的曼哈顿、芝加哥的中心区等，曾经造成了城市中心区的萧条，犯罪现象也有所上升，这些都是大规模的城市开发可能对城市环境带来的负面影响。

（5）项目具有明显的政策性色彩

城市大规模项目的上马大多是从社会发展的宏观战略出发，以推动国民经济发展和社会长治久安、人民安居乐业为最终目的，因此不论是否由政府直接参与，项目成败都会影响到现任政府的形象，是评判政府政绩的重要标准。另外，城市大规模项目涉及社会生活的方方面面，从决策、建设到运营，一直是社会舆论关注的焦点。社会舆论对城市大规模项目的影响是巨大的，正确的社会舆论导向是影响项目科学决策、顺利实施的重要因素。美国旧金山市海湾地区捷运系统（BART）计划就是由于居民要求保持独栋和低密度住宅群的继续存在而反对，最终不得不放弃该计划中两个车站的建设，至今仍未开发。

上海市部分城区的人口变化[2] 表1-1

地区	1995年		2005年		人口增减		人口密度变化
	人口数（万人）	人口密度（人/km²）	人口数（万人）	人口密度（人/km²）	（万人）	（±%）	（人/km²）
黄浦区[3]	76.92	124311	59.85	48227	-17.07	-22.19%	-76084
卢湾区	39.33	48856	31.67	39342	-7.66	-19.48%	-9514
徐汇区	80.62	14723	88.74	16205	8.12	10.07%	1482
长宁区	60.75	15862	61.84	16146	1.09	1.79%	284
静安区	40.47	53106	31.00	40682	-9.47	-23.40%	-12424
普陀区	82.57	15060	85.8	15648	3.23	3.91%	588
闸北区	67.01	23512	70.51	24098	3.5	5.22%	586
虹口区	82.7	35223	78.50	33433	-4.2	-5.08%	-1790
杨浦区	106.2	20373	108.16	17810	1.96	1.85%	-2563
浦东新区	148.63	2843	184.81	3535	36.18	24.34%	692
闵行区	54.4	1467	82.52	2220	28.12	51.69%	753
宝山区	69.18	1629	80.80	2982	11.62	16.80%	1353

---

[1] 空心化，最早指西方一些国家大城市由于市中心地价高昂，只能开发商业、写字楼等项目，造成白天熙熙攘攘、夜晚冷冷清清的现象，现在在发展中国家也有出现。
[2] 上海统计年鉴，2006，本研究整理。
[3] 南市区2000年已并入黄浦区，故2005年的数据已计入黄浦区。

上海市部分城区社会消费零售额变化表[1]（亿元）　　　　表1-2

地区	1996	1997	增减（±%）	增幅排序	1998	增减（±%）	增幅排序
黄浦区	129.49	130.39	0.70%	10	127.56	−2.17%	12
卢湾区	53.11	58.79	10.70%	5	60.20	2.40%	10
静安区	34.76	37.46	7.76%	8	46.51	24.16%	1
徐汇区	61.00	64.78	6.20%	9	70.22	8.40%	8
长宁区	42.76	48.70	13.90%	4	56.38	15.80%	3
虹口区	54.04	53.93	−0.20%	11	56.36	4.50%	9
普陀区	46.36	56.47	21.80%	2	64.57	14.35%	5
闸北区	54.17	58.67	8.30%	7	57.73	−1.60%	11
杨浦区	55.72	52.10	−6.50%	12	60.02	15.00%	4
浦东新区	140.22	162.23	15.70%	3	178.97	10.31%	7
宝山区	52.07	56.39	8.30%	6	62.50	10.80%	6
闵行区	49.90	64.57	29.40%	1	77.86	20.60%	2

（6）资金筹措复杂，随着投资建设主体及渠道的多元化，有些项目甚至不是由政府发起建设

鉴于当前的经济发展状况，无论是发达国家还是发展中国家，城市大规模开发项目都不可能全部由政府投资，项目资金的来源除国家投资外，还包括诸如国内外贷款、债券发行、项目受益地区及部门的集资、项目前期工程滚动开发的收入以及股票的发行等多种方式。随着投资建设主体及渠道的多元化，有些项目甚至不是由政府发起建设。只是对大规模项目来说，政府的参与十分必要，因为项目建设中不仅需要政策的许可与配套，同时也需要有来自政府的直接投资或融资担保。不同的投资来源使政府负担不同的经济风险、社会风险和政治风险。因此，政府不论是否为项目的发起人，不论直接投资与否，它都是大规模建设项目资金筹集的组织者和管理者。

（7）项目决策程序长而复杂

其决策过程均需要通过完整的可行性研究及环境影响分析、需求分析等比较后才能完成，国内如武汉等城市的轨道交通项目仅可行性研究的时间就超过10年。

（8）决策因素变数多，政策、政治、国内外形势，甚至国家间的政治与贸易关系会在决策时起到关键性的作用

通常可能由于政策改变、政治变迁、国际国内形势变化，而影响项目的决策结果。某些项目由于拟采用的技术方案、资金渠道等原因，国家间的政治与贸易关系会在决策时起到关键性的作用。如上海磁悬浮项目受到中德关系的影响，京沪高速铁路在磁悬浮、轮轨模式选择方面的争端涉及中德、中日关系。

（9）决策层均为高层负责人

由于项目规模大、牵涉方方面面的环节，对城市而言属于重大决策的一环。因此，通常需要由较高层的负责人进行最后的拍板决策，有些区域尺度的大规模项目甚至需要中央政府来统筹协调及决策。

（10）项目具有高知名度

按照规模可以对大规模项目进行分类，区域尺度、城市尺度以及地方尺度的项目（Borja and Castells, 1999）分别对应于机制上、管理上及操作上的层面，三种尺度中都具备通用的协议、职责和谈判结构，大规模项目可以在三种尺度中都有所体现。

---

[1] 上海年鉴，1999，本研究整理。

区域尺度的项目涉及整个地区在战略层面的问题，即战略协调（而不是过去几十年里多地方、多部门各自为政的干预）以及在城市网络中集聚的位置，是依据经济、社会、环境及文化目标而对未来理想状态的追求。

城市尺度的影响则针对城市化的区域，为了提升城市的全球地位、提高生产力及改善不同的活动节点间城市网络而做的策划。

地方尺度的干预指那些虽然规模小却有具体目标，在城市发展中具有战略功能的行动。它们被引导去改善形象、解决不足、改善多功能活动或其他基于城市现有动力的目标。地方性的战略干预从邻里复兴到新的商业街、旅游或娱乐设施的建设、历史性中心的重建等。

具体而言，城市大规模项目包括当前非常盛行的城市发展战略、基础设施、大学城、科技园、旧城改造、超大型活动（如奥运会、世博会）等，其作用包括改善城市形象、提供高科技办公空间、提供国际标准的基础设施和通信设施、提升交通可达性及流通性，力求提升城市对各种产品和服务的生产能力。具体方式包括赋予闲置土地以新的功能，改变生产用地的用途（工厂、港区等），加快城市现代化进程，建设地标性建筑，提供公共开放空间和绿地等。这些战略性的干预项目都需要调动公共及私人资源，要经过项目可行性、项目管理和影响的分析，能够起到改善城市形象与跨产业协调的作用，是城市营销计划的重要组成部分。

1.5 城市大规模项目的作用和参与者

1.5.1 土地调节

干预城市化进程是公共部门的习惯做法，只是由于全球化的影响，造成国家干预所扮演的角色、本质和原则改变、结构调整的不同而不同，不同国家因文化与社会经济背景的不同也呈现出差异。

在大规模城市项目中，公共部门可以是被动的角色，通过在合作合同中明确一定的条款来约束私人机构必须满足一定的社会及环境目标，以保护公众利益。

一方面，公共部门也可以是积极的主导角色，通过公共项目和财务机制（税收、信用、补贴等）的安排，以及区划和规定来实现干预。另一方面，有些国家城市土地被视为公众财富，国家有权为保护全社会的整体利益而管理土地。许多曾经的社会主义国家废除了私人可拥有不动产的政策，盈余分配的实施机制，例如土地没收、部分或全部城市土地的公共收购等，成为国家控制城市土地以干预城市化进程的主要做法。

许多研究证实，完全基于自由市场机制的城市发展存在着危险，并会带来很高的社会成本（Alquier，1971；Acher，1972；Clichevsky，2002；Lungo，2002；Smolka，2002）。用于土地上的私人投资或许是造成大都市"社会—空间"分隔的重要起因，还说明了公共行为与私人利润间的关联。同时，中央计划体制下的城市发展一直以来都无法达到其目标，显然，土地干预对帮助旧城区大量废弃、贫困、衰退的地区实现复兴有决定性的作用，特别是当发展中国家的土地管理致力于实现社会与环境目标的时候。

目前，城市大规模项目的实施再次提供了一种可能性，以思考如何控制土地来实现社会与环境目标。土地及房地产增值的实情，证明了市场化条件下房地产业者追求盈利的天性，也意味着价值的创造和私人拿走增值，以及城市大规模项目的存在不可避免。许多国家因此制定了相关政策，以将房地产中的增值用于实现社会或环境发展的目标，著名的例子包括美国、加拿大、巴西的法律允许实施"空中开发权"[①]、"土

[①] 空中开发权（Development Rights）是指土地上空的使用权利。就法律传统而言，美国法虽属英美法系，但其传统的包括不动产在内的财产法制却继承了罗马法，从而认为土地所有权的范围"上达天宇、下及地心"。进入20世纪20年代以后，由于城市人口急剧增加，美国进入了一个前所未有的城市土地的立体开发与利用时期。这一时期，转让、出租某一被规定上下范围的地上、地下空间，以获取经济利益的现象不断发生。这种以空中之一特定"断层"为客体而成立的权利，被称为"空间开发权"。

地再生权"①等交叉补贴方式。在许多国家，着眼于解决社会问题的市民运动、新出现的社会角色以及传媒通信业的发展已经在制定新的土地政策及合理利用开发获取土地增值的使用上发挥重要作用。"市民参与"则成为城市大规模项目决策过程中的一个重要环节。

由于国家财力有限，大规模项目因其规模而需要完整的可行性研究，以及清晰而透明的国家管理机制。这样，新的方式给予国家以主导权，即在吸引私有资金参与建设的框架下，在公共与私有部门以及社区间形成合资机构。城市建设所面临的新局势已经改变了干预的尺度，仅凭一方的力量无法实现城市结构的优化。新的城市战略试图找到一些能够像催化剂一样产生连锁影响的项目。大尺度的（或大量的）干预被认为是一个好的机会，因为它允许不同角色（公共或个人）在复杂运作的框架内发挥作用（Garay，2001）。传统的土地利用控制机制已经转变成一系列的程序以及制度、步骤与协议等手段，例如下面这些城市战略运营的典型案例：在纽约的贝特瑞公园城（Battery Park City）项目中的政府角色；在伦敦码头区（Docklands）和巴黎的集中重建区（ZAC）为取代严格的土地控制规定而使用的分散化土地供应方式；在巴塞罗那为了Vell港的重大事件而进行的一系列大规模建设项目；在圣保罗的Agua Branca和Faria Lima项目中的关联运作；以及波士顿的收费道路项目中的"空中开发权"的转移等。

在国外，城市大规模项目在很大程度上集中于数十年衰退的旧中心区复兴中，许多市场经济国家在城市发展中出现问题的原因之一是房地产类型的过度细分，因为土地及建筑用途的过度细分已经很难满足城市新的全球化功能要求。缺少这样的土地过去阻碍了商务功能继续留在城市中心区的可能性，还导致了逆城市化、中心区衰退、城市土地和建筑价值的下跌。逆城市化进程将投资者和开发商引向城市其他地区，并在城市边缘产生了新的中心和活动。与此同时，无所不包且快速影响着生产过程的科技进步，生产与消费的流通则改变了企业的区位选择，加上运输、通信方式及服务管理的改变，城市中出现了大量的废弃土地。

寻找一个更有效的土地管理和基础设施配置策略已成为城市适应变化的重中之重。土地政策把焦点集中在一系列目标上，其中一个优先目标就是恢复城市中心区的居住功能，这是在前几十年中明显减少的趋势，其他目标还有恢复旧城区及其自然环境、滨水区或其他景观，比如那些在城市扩张中丢失或者失去其韵味的景观。其他的目标还包括消除在发展中国家普遍存在的城市结构分裂和基础设施服务标准的不平等，以及不断上升的城市贫富不均现象。

得益于城市大规模项目的实施，土地政策及税收可以成为国家引导和调控以确保空间及社会目标实现的手段，国家通过区划（限定土地使用类型、强度、形态等）控制土地的能力直接影响到各个分区与建筑单体，并吸引开发商到衰退地区投资，然后通过税收等政策对土地与建筑开发中增值部分进行再分配。

土地政策通过区划条例"创造土地"的途径也有多种形式，如对中心区的复兴活动给予更高的容积率等。

巴西圣保罗市关联运作的例子是有关如何在一个土地和房地产价格持续增长时期建立公私合作关系（Public-Private-Participate，简称PPP）。1992年，公共资金的缺乏、城市政策的不足以及为社会问题获取城市化土地的困难，迫使当地政府为了解决社会住房和其他穷人生活的问题而与私营机构达成妥协。在关联交易的过程中，当地政府将超过城市规划规定的指标（容积率、建筑密度、建筑面积等）售于个人（图1-1）。

在波士顿的案例中，"空中开发权"与一个1997年的Turnpike收费道路项目联系在一起，牺牲掉部

---

① 土地再生权指根据公众利益的需要，允许通过修改土地使用用途而获得相应的开发收益。

图 1-1　圣保罗 Faria Lima 的城市运营项目（P. Sandeioni 摄）

分土地以及土地价值的增值部分，使原本遭到反对的 Turnpike 项目得以实施。在萧条时期征收土地，经济复苏后市政府可以利用手头的储备来满足住房、公共设施、研究设施和私人投资的需要。前面的例子中，市政府通过出卖空中权给马萨诸塞州收费公路管理当局，来自收费道路走廊下方的空中开发权转移发挥了重要作用。这一运作既建设了道路，也取得了需要的土地。与第一轮的城市现代化目的不同，这一轮的城市现代化重点在于鼓励混合利用土地，迅速应对不同活动间产生的冲突，灵活地修订规范，以允许土地功能和开发强度的调整（图 1-2、图 1-3、图 1-4）。

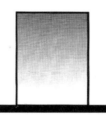

图 1-2　波士顿的收费道路项目，空中开发权在此发挥重要作用

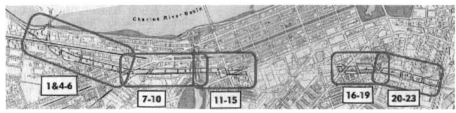

图 1-3　波士顿的收费道路项目，空中开发权在此发挥重要作用

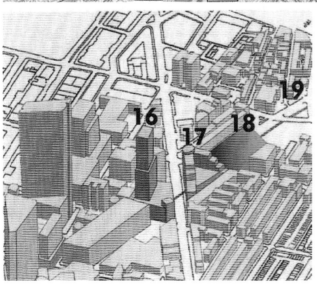

图 1-4　波士顿的收费道路项目，空中开发权在此发挥重要作用

图 1-5 海牙市中心以居住为主的大规模城市更新项目
注：图中 1-11 为各主要更新项目。

为适应全球经济发展的需要，特别是以出口为导向的金融、公关与广告服务、文化与旅游产业呈现出日益重要的趋势，需要调整原来规划的用地空间结构。在很多情况下，调整就是降低中心区制造业用地的比重，以适应中小型新企业的需要。

新的土地利用规划通过大规模项目加速了将废弃地、旧楼宇及设施更新改造的过程，新的用途包括：交通系统与枢纽、铁路、轨道交通、港口与集装箱码头、机场及空运设施、旱港、新科技园区、酒店、文化及体育设施、博物馆、会议中心、通信设施等。在新的土地利用规划中既要保持及扩大公共空间，同时还要维持相当的居住密度（图 1-5）。

1.5.2 改善管理的机制

要成功实现城市的转变与更新，在城市管制上需要采纳以协调为主的政策并意识到私有机构的参与以及使用者投资的重要性，使整个都市的局部地区在通过大规模项目实现快速城市化的同时，也可以满足全球化定位时城市多功能的需要。

虽然可达性、通信和流动性的改善对于容纳这些新的机会非常重要，但是很显然，透明度及强制性的条款也是实现这些变化必不可少的条件。发展中国家的城市重构表明了形成和加强这些新中心的必要性，同样重要的还有从空间上整合这些新的中心及其周围的广大地区。

与此同时，也必须注意到大量的公共资源用于少数特权区域的市场运行，会加剧设施与服务分配的不平等，对贫困区域带来严重的负面效果。

1.6 大规模项目成功的要素

结合对众多案例的分析，根据城市大规模项目的特点，可以对影响大规模项目的 29 条关键因素[①]汇总如下：

1）项目目标和前期策划（9 条）

（1）规划越早介入项目，规划中越尊重市场因素，越容易得到开发者的认可，项目成功的可能性就越大，规划也越可能得到实施而获得成功；

（2）市场需求与开发机构目标、项目规划与设计、项目建设及后续管理之间的一致是整体项目规划策划的前提；

（3）项目的目标定位要分阶段、分层次，并考虑到未来的动态变化；

（4）项目实施过程中，城市交通调整、财政安排等客观制约条件要予以充分考虑；

（5）参与项目建设各方的态度及其之间利益分配的协调很重要；

（6）要有必要的事实或理论依据来全面定义项目目标，目标表达清楚；

（7）在项目可行性分析时，充分估计政府领导人更替、宏观政策变化、市场变化等政治及资金因素对项目目标实施的影响，并考虑适当的应变措施，即对政治风险、政策风险、市场风险要予以充分预计；

（8）充分重视社会舆论及媒体对项目的影响；

---

① 马文军. 城市开发策划[M]. 北京：中国建筑工业出版社，2005：39-41.

（9）注意防范规划、决策、策划失误可能带来的负面影响，并制定防范措施。

2）政治、社会、环境因素（9条）

（1）政治、社会、生态环境及其他项目外部环境因素从根本上决定了项目的成败；

（2）项目立项前，应认真分析与项目有关的社会经济、政治状况以及社会群体的心态；

（3）项目建设需要与城市中长远发展规划相适应；

（4）城市大规模公共开发项目的成功实施，需要创造品质良好、多样化、美观、温馨、欢愉的城市空间，这样才能吸引更多的人群使用项目提供的产品与服务，才能吸引更多的投资商参与到城市开发建设的投资活动中来；

（5）如果政府与企业或私人联合投资，意味着项目目标和利益倾向的多样性，为后期实施带来困难；

（6）制订行政保障和立法保障措施以配合项目的顺利进行；

（7）妥善处理、协调好项目与社会群体的关系，在各阶段均做好公众参与工作；

（8）协调好各级政府的关系，即项目与所在区域的关系；

（9）在涉及外资的项目中，要考虑国际政治关系、经济关系以及民族感情因素的变化。

3）项目投资（4条）

（1）及时跟踪、分析项目实施背景的变化对项目目标、投资的影响；

（2）吸引对项目本身（社会效益）有兴趣而不仅仅关注项目投资回报率的投资商；

（3）在项目的投资风险分析中，要明确担保行为和政府担保措施、业主责任等，同时考虑到通货膨胀因素，充分估计汇率的变动；

（4）资金来源要稳定，并做好必要时中止项目投资的准备，即项目停工等待。

4）项目设计和技术（7条）

（1）面临重大的社会或经济变动时期的项目设计方案中需要准备必要的多种方案与替代措施，以应付不同情况的出现；

（2）涉及项目定位的重大前提应该保持持续性和稳定性，但一旦有所变化，规划与设计必须迅速调整以配合；

（3）在设计方案未作最后定论前，认真全面检查设计方案；

（4）加强设计阶段的管理，在项目的各个阶段，慎重对待设计的更改；

（5）认真做好有明显依存关系的技术因素的协调工作；

（6）不断地修改方案的结果可能使原有设计亮点全无踪影，因此设计一旦完成，就应"冻结"设计工作，以保障设计者初始构思的闪光点不被轻易地抹掉；

（7）每一细节的设计错误，都将有可能影响到项目的投资和进度。

因此，在大规模项目的规划阶段必须抓住这些因素，在充分了解市场需求的基础上加以规划整合，形成对大规模项目建设的指导纲领。同样，政府的规划管理机构要以策划的观念与方法，谋划本地区的战略发展，并对地区内的建设项目加以正确的引导（图1-6）。

1.7 结论

首先，信息与通信科技的持续发展、生产与文化加工上的非实体化、国际资金流动的加速，导致了城市风格的复兴，以及对城市空间的重新定义。全球化的环境为大规模项目的实施创造了机会，投资方式的改变也重新界定了政府对于城市空间组织的作用，从直接的创造者转变为促成空间虚拟化的推动者。这一变化推动了城市内部的结构变化，不论发达城市还是欠发达城市，社会不公平现象和自然环境的脆弱性日益加重。

其次，第一轮的城市现代化影响到城市的整体结构，提出了建筑空间与开放空间的关系，而第二轮的

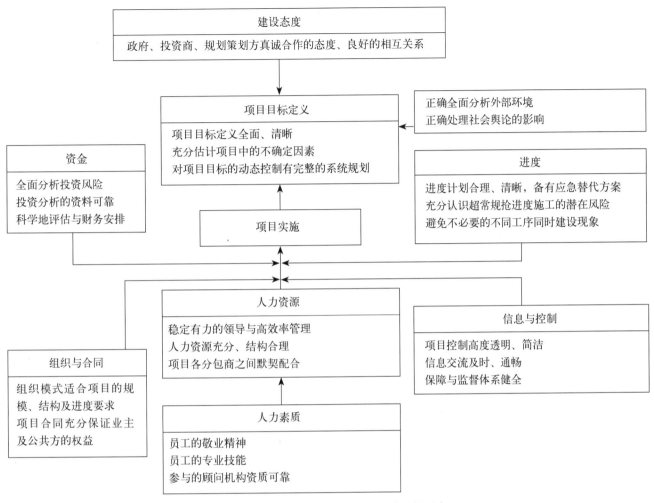

图 1-6 城市大规模开发项目成功的关键因素

城市现代化改变了城市的一些局部地区，影响到公共空间与私有空间的利用与关系。第一轮现代化考虑到大城市中心的疏解、交通对于城市结构的重要作用、绿化带在城市内部整体结构中的作用、步行与机动车交通的区隔、传统街道与现代交通干道的区别。第二轮现代化中关注的主要是城市在国家发展战略中的作用，因此，利用聚集形成的相对优势，营销城市及提高城市生产率是战略规划中重点考虑的问题。

第三，虽然城市的发展因工业化、投资而推动，但控制土地的政策如城市规划、条例、居住法规与公共空间乏力的作用更加重要，以调整社会收入的再分配和保护环境，因为即使是最黑暗的剥削统治时期也兴建了城市广场、通道和其他必要的公共空间。

第四，一方面，城市大规模项目是一种使城市区域特色同质化的社会文化冲击，特大城市的出现、城市地带的分裂以及社会经济行为的多元化是发达国家及发展中国家共同的特点。另一方面，由于公共资源集中于大规模项目，可能造成其他城区缺少经济住宅、安全及最低生活保障，社会排斥、交通拥挤、环境污染、正规经济萎缩也难以避免。越来越需要更加有效的土地政策和基础设施配置方式，避免贫富分化可能造成的社会分隔，提高城市生产率和改善低收入群体的生活。

最后，城市大规模项目的成功实施涉及 29 个关键性要素，需要在准备阶段逐一应对，城市规划要起到策划的作用，从战略准备、社会公平、环境保护等方面把握项目的建设目标，成为全球化背景下城市创造竞争优势的指导性纲领。

第2章 经济危机冲击下的城市发展
——营销导向、"精致城市"及其他

2008年，美国次级贷款危机引发了全球金融风暴，导致各国经济出现不同程度衰退或增速放缓。在危机的发源地美国，大批金融机构破产，三大汽车厂商陷入危机，失业率连创新高，消费市场萎缩；欧洲各国也陷入了严重的经济危机，最为严重的当属冰岛，这个一度全球最富国家之一的国家竟面临国家破产；金融业发达的英国也出现了经济衰退，自1992年以来，经济增长首次出现萎缩；由于全球金融危机以及海外需求疲弱，日本经济陷入严重的衰退，2008年第四季度经济创1974年以来最快萎缩速度，GDP季比下降3.3%，折合成年率下滑12.7%。

面对金融危机各国纷纷出台措施应对，美国推出多轮经济刺激方案，包括注资金融机构、减息降税、刺激就业等等；英国出台包括就业刺激和信贷担保等内容经济刺激方案；法国相继出台了银行业、汽车等救助计划。

由于我国经济外贸依存度高达60%，而中国出口的主要对象美国和欧洲的经济萎靡，造成我国出口大幅下降，进而造成我国经济增长速度下滑。2008年中国的经济增长率只有9%，是7年来的最低水平。

为抵御国际经济环境的不利影响，我国出台了扩大内需的十项措施，以加快民生工程、基础设施、生态环境建设和汶川地震等灾后重建的速度，提高城乡居民特别是低收入群体的收入水平，促进经济平稳较快增长，并先后推出钢铁、汽车、船舶等十大产业振兴规划。全国各地方政府也相继推出了扩大内需的措施，如北京、上海、深圳等等。经过两年多持续的建设，我国不仅再度实现两位数的经济增长，还在整个国际经济动荡中发挥了重要的稳定性作用。

然而，造成2008年国际性经济危机的原因并未根本消除，尤其是以美国为首的发达国家，奉行了依靠新兴经济国家廉价产品贪婪超前消费，同时放松金融监管，放任华尔街精英们以多种方式攫取利益，造成实质性经济增长速度远远低于开支的增长，从而形成债务不断增加，偿债能力不断下降，不仅希腊、西班牙、意大利等传统发达国家陷入债务危机，连美国作为全球第一经济大国也被品级公司标准普尔调低了国家主权信用评级，经济衰退的阴影笼罩全球。

2.1 危机中的城市发展机遇

在这次全球性经济危机的冲击下，发展中国家广受波及，也暴露出近些年以来中国经济增长过分依赖出口，特别是低附加值、高能耗与资源消耗的制造业产品出口，而对地方政府的考核中过分关注GDP的增长数，导致各地方政府一切围绕提高GDP增速的指挥棒，纷纷招商引资上项目，并将主要资金投入尽可能用于能提高产值的生产性项目建设，城市基础设施和民生工程的投入相对不足，从而影响到城市的健康发展，城市建设方面欠账很多，城市发展呈现粗放型增长态势，多数城市环境品质不高。2011年夏天以来，北京、武汉、成都、上海等特大城市相继遭遇了大雨等灾害的袭击，城市的生命线系统遭遇严重威胁。

分析造成这些局面的原因，如果说过去我们的企业主要关注销售量而忽视品质与品牌的取向，使他们关注较多的是城市提供的各种优惠政策、地价等硬性的指标，对于城市各方面软实力的关注不太多，因此城市对这些方面也不够关注。而当前，企业已经开始意识到必须通过创新以提升产品品质，从而获取竞争优势和实现转型，而品质提升需要的人力资源等依托必须依靠所在城市提供。

如此看来，作为企业发展的舞台，城市既面临挑战，也存在机遇。一方面，国家推行扩大内需的政策后，城市可以通过相关项目的建设，合理引导政府投资，提升城市品质，提高居民生活水平，完善城市的基础设施建设，为以前城市建设的相对滞后补课，实现城市的转型。同时，政府的众多投入，需要科学的策划和管理水平，才能够做到高效率、高效益，发挥推动城市发展的作用，因此对城市规划就提出了很高的要求。

2.2 城市规划视角下的内需与城市诊断

为了抵御国际性经济危机的冲击，调整国家经济增长过分依赖出口的倾向，国家提出了扩大内需拉动经济发展、调整结构转变经济发展方式的政策。这里说的"内需"是指国内需求，包括投资需求和消费需求两个方面。扩大内需，就是通过国内基本建设投资加大，在国内资金充分的前提下，优先加强基础设施的建设，并利用基本建设投资，带动经济增长；同时，提高人民生活水平和工资水平，使其有花钱购买的能力，特别是通过市场引导，引导居民将钱花在固定资产投资上。通过投资市场与消费市场的启动，拉动整体经济增长。这里的"调整结构转变经济发展方式"，不是不要GDP，而是要健康的绿色GDP。

而从城市规划视角看，内需是指对城市发展起到支撑和推动作用、城市发展必须的内部需求，包括城市的基础设施、公共服务设施、住房、绿化等城市硬实力要素建设，以及文化、科技、教育等多种城市软实力要素建设。城市通过上述内需的建设，一方面能够提高城市公共服务水平，为城市的长远发展奠定基础，另一方能够增加就业，提高居民收入，从而提高居民的消费能力，达到拉动经济发展的目的。

要扩大城市"内需"，就要将城市健康发展所必需的内需摸清楚，通过诊断城市目前各种设施的水平，分析城市发展的需求（包括保障性住房的需求、基础设施的需求、城市环境建设的需求等等），以及城市拥有和能够调动的资源，做到了解城市发展的制约因素，明确城市发展的正确方向，引导有限的政府资源用到效益更高的地方，这就是我们常说的，合理有效地配置资源。

然后要制定近期行动规划。在城市诊断的基础上，确定城市发展所急需的项目，根据项目的重要性和迫切性，考虑城市财力的约束，科学进行项目的论证，安排各种工程举措，改善交通条件，提高城市运行效率；建设保障性住房，改善居民的居住条件；加强城市能源、水、电、环卫等基础设施的建设，从而改善民生，提升城市品质。

此外，还要制定长期行动规划。"人无远虑必有近忧"，通常一届政府最关心的是自己任期内的事情，而城市的发展需要有长期的规划来指导，很多对城市发展具有重大战略意义的大规模项目（Large Urban Project）很难在短期内完成，比如一些大型基础设施建设、城市品牌打造、人文环境塑造等都非一朝一夕之功能够完成的。这就要求政府具有长远眼光和宏观战略，城市规划要为城市的长远发展进行空间布局与设施安排，制定未来城市发展的蓝图，为城市的未来蓄势。

2.3 城市规划与策划城市发展

2.3.1 城市规划的职责

城市规划的直接职责是贯彻城市的发展方针，合理安排城市规划区内产业的区域布局、建筑物的区域布局、道路及运输设施的设置等工程，通过对各种具体设施的安排而直接策动城市发展。

城市规划的间接职责则是协调城市内部各方面的关系，为城市的发展提供有力的支撑，减少各种活动的负面外部性，优化城市系统的运行，主要侧重于内部关系的协调，以及用地和空间的保障。

这样，城市规划中的规划与设计就像企业产品的设计过程，主要围绕有形的建筑物与环境来做，而城市策划就像与该产品营销相关的市场调查分析、产品定位与市场细分、用后评估等相关工作。如果我们把城市内的居民与期望吸引的外来投资者、消费者比作用户，而把城市及其提供的环境与空间比作产品，城市规划策划就是针对城市规划中属于策划范畴的那部分工作而进行的谋划活动，也就是对城市发展的策划。在具体实践中，要根据战略、管理及项目等不同层次规划的需求，针对各种城市问题进行目标选择，并寻找实现目标的对策与途径等。

2.3.2 经济危机冲击下，我国城市面临的困境与规划策划

全球性的经济危机影响着欧美日本，并波及我国及其他发展中国家，又从企业界逐渐扩大至社会各界，各城市的领导者都面临了社会就业、企业生存、财政收入增长等问题的困扰，并各自寻找着适合自己城市的对策。

众所周知，浙江义乌是一个以小商品市场闻名全国乃至全球的县级城市，通过"兴商建市"创造了经济发展的奇迹，形成的"义乌模式"对全国很多地方的发展影响很大。然而随着市场竞争的激烈和外部环境的变化，义乌发展面临越来越多的困难，这与全国很多地方都开始兴建商品市场，争夺了义乌的市场空间不无关系。因为义乌是一个县级城市，本身内部市场不大，空间狭小，竞争中缺乏优势。

义乌的领导人认识到义乌的优势在于先发地位和管理经验，只有走出去，才能打开新的局面。这方面值得借鉴的企业案例很多，比较成功的如上海的"新天地"项目，该项目以中西融合、新旧结合为基调，将上海传统的石库门里弄与充满现代感的新建筑融为一体，集餐饮、购物、娱乐等功能于一身，成为国际化休闲、文化、娱乐中心。伴随上海新天地的成功，其开发者瑞安集团先后推出了西湖天地、重庆天地、汉口天地等项目，更有其他城市正规划建设与上海新天地类似的项目。义乌学习了新天地的经验，发挥自己的优势，输出管理，在全国各地规划集中展示基地，塑造城市形象，推广本地"产品"。

福建晋江是另一个经济高速发展的城市，产品品牌建设走在同类城市前列，被冠以"品牌之都"的美誉。对于当前出现的危机可能造成的影响，当地在土地与项目摸底后，规划遴选合适的地块与项目作为推进城市建设、改善城市面貌的措施，并在城市机场旁规划、策划了本地名牌展示中心，试图向往来城市的高端乘客推广本地优质的产品品牌、环境优美的商务休闲环境和精致的城市"目的地"。

浙江温州在发展中也遇到了与义乌相类似的问题，温州因为企业的灵活经营、敢于冒险、善于把握市场动向而占据先机，然而温州企业也在快速的扩张过程中给国内外用户留下过假冒伪劣的形象，最近虽有所改观，但很多中小企业欲独力打造产品的精品形象还很困难，特别是当前外贸出口的急剧下降使他们对外拓展的步伐严重受挫，因此温州市政府及行业协会拟牵头在全国大中城市建100个名品展示中心，集中展示推介本地产品与城市。通过在外展示本地高品质产品，改善大家对温州产品的认识，一方面提高温州产品的声誉，同时也能塑造温州城市的高品质形象。

2.3.3 城市策划的层次

城市规划承担着策划城市发展的职能，对应着城市在战略、管理及项目层次的需要而分为城市战略策

划、城市管理策划及城市项目策划。

战略层次的城市战略策划，指对全局性、高层次的重大问题的筹划与指导，针对城市领导者所思所想的城市战略问题。通过调查研究，汇集多方思想，明确关乎城市未来的重大目标，制定备选行动方案，为城市重大事件提供全局情报咨询、信息汇总、决策方案和实施规划，并为总体规划服务。

管理层次的城市管理策划，指的是为实现整体的长远发展目标，在城市中观层次上细化总体目标的实现举措，并据此明确各种规划制定的原则，是为完善日常规划管理而进行的策划，也是对规划的"规划"。

项目层次的城市项目策划，指的是针对上一层次策划明确的某个具体项目，在规划、操作过程中具体化、对象化的策划，可以是包含了创意、策划、战略、谋划与设计的行动方案。

三种层次的策划对应于城市发展的不同需求，也对应于城市规划的不同阶段与职能，用以完善城市规划工作，满足包括城市领导者、城市规划管理者以及规划设计工作者等不同人群的需要。

2.4 经济危机时期的城市应对

在信息高速传播、资源快速流动的今天，市场经济已将城市逼迫到竞争的环境中，商业社会则使城市不得不面对被选择和被抛弃的压力，要求城市随时保有明确的发展目标和策略。同时，在科学发展观指导下的国家经济发展模式，将从粗放型增长向集约型、创新型转变。而随着经济社会发展和人民生活水平的提高，消费需求将替代投资拉动成为经济发展的主导力量，特别是内需。

过去中国城市发展更多的是实行成本领先战略，通过土地、人力的低价格来吸引投资，进而做大GDP总量。在宏观形势好的时候，品质高和品质低的城市都能获得发展，只是快慢有所不同。然而，当危机袭来，城市发展竞争变得激烈，城市自身的品质变得极为重要，当企业开始选择城市，那些品质低的城市发展就会陷入困境。现在我国很多城市开始出现企业生存困难、出口下降、就业困难的问题，以前的发展模式已经难以为继。

迈克尔·波特在他的竞争理论中认为，国家和区域的竞争优势要有动态性，也就是在以前竞争优势不复存在时，需要寻找到新的优势，优势才能长久保持，作为城市也要在面临困境时选择新的发展道路。那些品质高、具备内生发展能力的城市能够持续发展、脱颖而出，在城市的体系中抢占有利位置，调动和享用更多的资源。在个人消费需求不断升级、企业不断升级的背景下，城市发展也要升级，脱离低水平发展的道路，建设精致城市，顺应城市发展由低级向高级的发展潮流。

在这一背景下分析，企业向社会提供的是产品与服务，并以其产品的高质量与精致可爱来获得好声誉，如人们趋之若鹜的奔驰、宝马的名车，LV、Cartier的名包、首饰一样，它们具有卓越的品质、传承性与历史，给人以美学与感官刺激。城市也是一样，除了城市内各企业提供的产品外，还有城市提供的"功能性产品"——空间品质和生活环境，由它给人们提供的身心享受而获得声誉。如果说，当前企业适时地调整企业战略，实施产品升级和品牌提升，以在金融和经济危机来临之际得以生存，并期冀有所发展，那么，在国家扩大内需与调整结构的政策环境下，城市应当避免过去粗放的增长模式，而要以提供品质生活、建设精致城市为目标。

2.4.1 精致城市

"精致"一词在汉语词典中有以下几层意思：①精细周密；②精美工巧；③美好，多用于对具体事物的描述，也引申至一种求美、向善的人生态度，是追求卓越的意志品格。

对于城市发展的目标来说，精致城市则是更加复

表 2-1 精致城市的含义

精致城市	物质层面	规划	它应该是一个紧凑型城市，是一个用地集约、规划布局合理的城市
		建筑	它应该是一个具有很多精致建筑、公共设施、环境的城市
		竞争力	它应该是一个在经济、文化、科技等方面都具有很强竞争力的城市
	精神层面	文化	它应该是一个拥有浓厚文化底蕴，而且较为完好地保存历史并适度利用的城市
		生活	它应该处于一种很舒适、很有格调的状态，应该具有自己的个性与品位

杂的界定。不同社会阶层、不同价值观的人会有不同回答，但能够一致的观点是，精致的城市一定是宜人的城市，需要有不可复制的文化内涵、生态健康的宜居环境以及良好的发展前景。具体来说，它不仅仅强调城市建筑的细部，也不单单指城市生活的舒适闲暇，它应该包括以下几种要素：

人们需要精致的城市与宜人的环境，而中国目前还缺少如同奢侈品一样精致的城市，正因为如此，少数特色鲜明的城市如丽江、平遥，已从大多数缺少风格与特色、没有准确定位和目标受众的芸芸众城中脱颖而出，成为人们心目中向往的旅游胜地。

2.4.2 精致城市的支持体系

精致城市具有独特性、艺术性、品质高和信誉佳的特性。尽管城市是一个复杂的系统，具有难以预测和控制的特性，我们还是要建立精致城市的支持体系，通过分解精致城市的目标，将抽象概念落实为具体可控的要素，把抽象的概念转化为现实。通过具体化和量化精致城市包含的内容，进一步建立可度量的指标，从而可以评价自身的发展程度，寻找自身不足，做到有的放矢。

2.4.3 拉动内需的城市规划应对举措——崇尚品质环境，建设精致城市

为了打造精致城市，城市规划应加强自身的策划能力，确定精致城市的支持因素，挖掘自身优势资源，弥补自身的不足，塑造城市品牌。精致城市的建设是一项系统工程，需要城市精雕细琢，全面关注，才能使精致城市的目标变为现实。

1）关注民生，提高居民住房和公共服务设施水平

生活在精致城市的居民应当享受高品质的生活，高品质的生活是城市发展的终极目的，而居民享受了高品质的生活才能充满自豪感，才能奠定城市声誉的重要基础。住房是居民日常生活的重要方面，

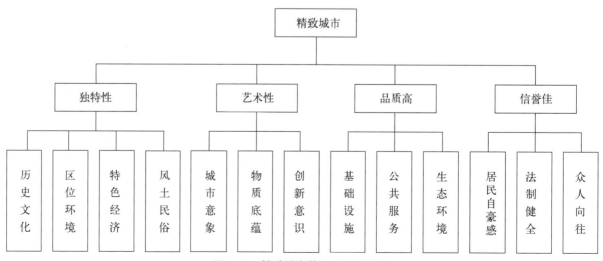

图 2-1 精致城市的主要支持体系

由于住房价格上升很快，目前有相当一部分群众住房困难，或者住房消费支出过高。据统计，近几年城镇居民收入中有超过 1/4 甚至 1/3 被用于支付住房消费。在这种情况下，住房消费严重挤压了其他消费，居民生活压力巨大，在这种压力下的居民生活难言高品质。国家扩大内需的十项措施中的第一条，就是加快建设保障性安居工程，城市应当多渠道筹措住房保障资金，合理制定保障标准，努力提高保障比例。学校、医院、图书馆、剧院、公园等公共服务设施同样是人们的生活水平不断提高的保障和体现，高品质的城市生活需要这些公共服务设施的服务水平随着社会发展而提高。

2) 关注长远发展，加强城市基础设施建设

过去很多城市政府在经济发展方面直接干预过多，也造成了公共财政对城市基础设施投入的不足，而在有限的财政投入中则更重视看得见的形象工程，忽视一些地下的基础设施建设，造成地下管网建设不足。据联合国有关组织的研究和建议，发展中国家的城市基础设施建设投资比例，一般应维持在同期国内生产总值的 3%～5% 水平上，或者为固定资产投资的 10%～15%，这样才能确保基础设施建设与经济增长的需求相协调。而我国城市基础设施投资在全社会固定资产投资和国内生产总值中分别所占的比重离上述比例都还有较大差距。1990～2003 年，中国城市建设固定资产投资占同期全社会固定资产投资的比重年均为 4.84%，最高年份也仅 8%，尚未达到联合国推荐的下限；占同期国内生产总值的比重年均为 1.78%，最高年份为 3.82%，仅略高于联合国推荐的下限值。因此，在当前国家加大投入的情况下，应该适时摸清本地现状，科学预测未来发展需求，合理制定建设标准，安排建设时序，补足交通、水、通信、能源、绿化等方面的历史欠账，为城市的长远发展奠定基础。同时还要避免基础设施建设过度超前，建成以后长期不能有效使用，挤占国家有限的投资，造成资源浪费。

3) 重视生态建设，建设高品质空间

目前我国很多城市是在重复西方发达国家先污染、后治理的发展模式，片面追求 GDP 总量的增长，对污染项目审批不够严格，对污染的治理投入不足，自然生态破坏严重，危害可持续发展。建设良好城市生态环境的基本要求就是要治理污染，污染严重的城市是谈不上精致的。要提高排放污水处理率，治理水体污染；合理布局工业，提高排放标准，改善生产工艺，治理大气污染；通过"限塑令"这样的管理规定减少固体垃圾的产生，通过分类提高垃圾的回收利用率，建立完善的垃圾回收网络，提高城市固体垃圾的处理水平；还要控制噪音污染和光污染等。

同时，精致不等于大，城市精致与否和城市规模的大小亦无直接关联。特别是在我国用地有限的国情下，城市还要努力提高开放空间的品质，在面积有限的用地上建设精致的公共空间。

4) 加强城市设计，提高城市审美素质和品味

精致城市不仅仅能满足功能上的要求，同时还要给人审美上的享受。它应当是一件艺术品，令人回味无穷，留下难以磨灭的独特印象，产生心灵吸引力。城市设计作为对城市空间的控制和指导手段，通过各个层次的设计，努力创造美好宜人的物质环境，体现城市地域特色和文化氛围，塑造城市良好形象。合理布局，科学组织城市空间，形成空间美感；通过对水体、山脉、绿地进行景观设计，营造环境美感；协调建筑设施个体、群体、总体之间的关系，使其友好对话，和谐共存，塑造整体美感；选择创意独特、设计精湛的路灯、邮筒、路障、消防栓、电话亭、公交站等公共设施，突出细节美感。

5) 抓住具体细节，从细处关注城市品质

有一本非常流行的书叫做《细节决定成败》，作者以大量案例论述了"细节"在管理中的重要性。这本书意在提示企业乃至社会各界：精细化管理时代已经到来。对于一个城市来讲，细节同样能够决定一个

城市的品质，追求精致的环境、精致的产品、精致的生活已经成为当前的趋势。城市规划往往关注城市的宏观发展问题，大型项目的建设，而忽视了一些细节问题，让我们的城市与一些发达成熟的城市相比，高楼大厦不少，但总是有粗糙的感觉。要建设精致城市，规划要更多地关注人性的基本需求，关注城市的无障碍设施、公共厕所、公交候车站点、信息服务、免费租车等等，从细节上提高城市的服务水准，既方便了居民，也能为城市赢得声誉。

6）推进产业升级，培育特色经济

城市直接提供的产品和服务能够给人带来直接的感受，优质的产品和服务能够提高城市的形象，巴黎的优雅、日内瓦的美丽、伦敦的历史都为各自赢得了声誉。城市的企业提供的产品和服务亦与城市形象息息相关，巴黎的时尚业、日内瓦的会展业、伦敦的金融服务业都为各自吸引了人流。一个泛滥假冒伪劣产品的城市是无法与精致城市的概念沾边的，城市品质的提升同样要求升级城市的产业水平，提高产品和服务的品质，形成几个代表城市特色的产业。产业升级并不意味着只能生产高科技的产品，高科技仅仅是一个方面，我们要升级的是产品的价值，通过流程升级、产品升级、功能升级和部门升级，谋得市场竞争优势，给城市发展带来持久的动力。

7）营销导向，打造城市个性品牌

现代营销之父菲利普·科特勒认为任何一个国家、都市甚至城镇都可以进行"场所营销"，鼓励城市结合自身实际建设，提升居住、旅游和投资的竞争优势和品牌形象。现在我国许多地方政府也有了城市营销的意识，热衷于举办会展活动、广告宣传来提高本地的知名度。虽然这些对一个城市来讲是正确的，然而这些宣传需要由城市的真实品质作为支撑，否则即使获得的城市美誉也将是无源之水、无米之炊，难以持续。杭州塑造"休闲之都"的品牌，根植于充分挖掘自己以西湖、西溪为代表的风景资源；武汉打造"光谷"，是依靠科技与教育优势，建中国最大的光纤产业基地。一个城市应当通过自己的特色历史文化、特色生活环境、特色经济产业、特色社会氛围来塑造一个城市的品牌。城市规划工作者应当努力挖掘城市本地的资源，准确定位，通过专业而合理的规划策划，制定城市整体发展与品牌创造的行动方案，协调不同部门的行动，打造城市品牌。

8）关注城市整体，致力吸引人才

城市本身是完整的，一种充斥着各样机能的有机个体。它可以比拟为人体的身体一般，是个活生生的主体，有着骨架、静脉及动脉。城市会呼吸、喝水、进食以及排放废弃物等；它会成长、再生与繁衍；它能思考并做出反应，有所谓的意志及精神。好的城市，如同人一般，其身心健康、有生气、有修养，并且有自信。好的城市具备值得称赞的模范特质和迷人的风采。就如同人类一般，城市有其不可缺少的活力干劲与驱动力量，需要满足维持其活力的机能。一个城市若要成功，不但在实质上是健康的，另外在文化、经济、政治与社会上，也一样需要是健康的。

由时间、区位的序列所构成的文化特质与城市纹理，形成了城市各自不同的特征，不只符合了各自物质上的需求，同时也表达其人民的信念与理想。这些特征的城市规划意义，简单地说，就是吸引人才的整体机会。

香港的品牌定位是活力与创新的"亚洲国际都会"，靠的是亚洲国际金融中心的地位和强大国际化的服务业、高标准的物流和电讯设施，以及掌握专门知识和技术的人才优势。而给这些人才以信心停留于此的，是香港所具有的多元文化吸引力，是国际水平的教育和培训制度，是国际品质的生存服务空间，以及对未来发展可持续的承诺。

9）持续改进，全面提升城市品质

戴明（W. Edwards. Deming）博士是世界著名的质量管理专家，他因对世界质量管理发展做出的卓越贡献而享誉全球。他的管理思想在战后日本工业振兴中发挥了重大作用,在他的思想影响下,索尼、松下等企业经过不懈的努力,成功推出具有优良品质的产品,发展为以质量闻名的企业。城市事务较普通企业复杂得多,精致城市的建设是一项复杂的工程,城市规划应当汲取戴明质量管理的思想,按照"PDCA戴明循环"的模式持续改进,不断提升城市的品质。PDCA循环是能使任何一项活动有效进行的一种合乎逻辑的工作程序,特别是在质量管理中得到了广泛的应用(图2-2)。P、D、C、A的意义如下:

P(Plan)——计划。包括方针和目标的确定以及活动计划的制定;

D(Do)——执行。执行就是具体运作,实现计划中的内容;

C(Check)——检查。就是要总结执行计划的结果,分清哪些对了,哪些错了,明确效果,找出问题;

A(Action)——行动(或处理)。对总结检查的结果进行处理,成功的经验加以肯定,并予以标准化,或制定作业指导书,便于以后工作时遵循;对于失败的教训也要总结,以免重现。对于没有解决的问题,应提给下一个PDCA循环中去解决。

10)加强规划策划,赢得全面竞争

生产出让市民满意并愿意消费与投资的产品,是城市持续经营与发展的唯一方向,同时也需要运用一个良好的营销策略去实施,所以,建立在营销导向基础上的城市规划策划是各个城市发展成功与否的关键。

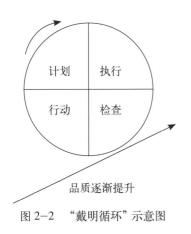

图2-2 "戴明循环"示意图

回顾20世纪的发展,借助铁路、飞机、货船以及通信的网络扩展,各国经济主体已经紧密联系在一起。全球化运动带动全球贸易成长20倍,总产值增加5倍,几乎没有国家能够抗拒加入世界贸易组织,加上网络交易活动的频繁与国际资金的快速流动,整个世纪可说是全球化由茁壮至喧嚣的一个世纪,跨国企业的资金到处流动,寻找适合的场所驻留,在获取利润的同时,也带动了所在地城市与国家的发展。在这个过程中,全球的市场与资金取向成了国家与城市关注的焦点,特别是一些特大城市,其扮演的角色已不再是过去那样简单,所要面对的竞争也不再只是局限于国内对手,必须具备多样化的机能与特色,才能获得国际性的竞争优势,占据领先地位。例如上海的国际化进程就要求其选择多样化的发展,以具有国际性的影响力,实现综合性的全球竞争优势。

全球化带来发展的同时,也带来了财富分配的不均,国际化企业和新科技产业的引进不仅带来高收入、本地化的管理与技术新贵,也导致低素质的劳动力大量被裁员,造成劳动力向高科技与低技术两极化聚集,使得族群分隔逐渐明显,而阶层间发生冲突的可能性开始出现。城市面对这一经济新秩序的改变,城市原有文化特征也在迅速改变。此外,科技产业在创造经济繁荣,改善人们物质生活的同时,新的污染也伴随而来,对城市环境的可持续发展提出挑战。

处于这样变化纷繁的时代,全球环境变迁已是不争的事实,城市面临的问题也日渐繁复,贫富不均、高失业率、社会阶层对立与环境失衡等等景象一一浮现,社会结构的失衡与冲突的新来源均显现出全球化带来的除去繁荣与国际文化多样性以外的真实风貌。因此未来打造新世纪的城市新秩序,需要从城市文化与精神、科技产业创新以及生态环境三方面进行思考,赢得时代环境的全面竞争与挑战。

因此,精致城市的创建不可能一蹴而就,需要长期坚定而持续的努力。同时,精致城市的建设也是一个动态的发展过程，不同生产力水平下有不同的精致城市标准与要求。因为人们的需求本身就是随着经济社会的不断发展而由低级到高级，由粗犷到精致。只有以科学发展观为统领，强调持续发展，真正从人的需求出发，"以人为本"，规划定位准确，标准建设得当，行动整体协调，才是形成优良城市品质与形象的关键，才能建设适合当代人需要的精致城市。

2.5 结语

"对于一艘没有航向的船来说，任何方向的风都是逆风"。城市的发展也不能漫无目的，"没有明确的战略方向，任何战术都无所谓好坏"，一个城市和区域要在21世纪空前激烈的竞争中脱颖而出，需要强大的城市竞争力，必须看清时势的发展，找准全球化大背景下新的城市战略定位。任何一个城市的超常规发展都是因为先行者有意无意踩准了时代浪潮的节拍，从20世纪80年代的深圳崛起，到90年代上海浦东的巨变，从沿海地区的天津、青岛，到中西部的重庆、成都，这些年的巨变并不是空穴来风，一定是从某种意义上符合了时代的发展规律，这些巨变预示着用全新的理念和思维方式打造城市新时代的来临。

在目前的形势下，企业如果因循从前的发展道路，只能在经济危机中愈陷愈深，城市与国家也是如此。城市发展面临着转型，城市必须找到新的方向，才能实现可持续的发展，精致应该就是城市发展的方向。为了向着精致城市的方向发展，城市规划要有新的关注点，要有新的理论与方法作为支撑。当前，国家应对危机和结构转型的政策既为我们指明了方向，同时也为城市规划的创新提供了机遇。

参考文献

[1] 马文军. 城市开发策划 [M]. 北京：中国建筑工业出版社，2005.

[2] 马文军，M. Carmona. 全球化与大规模城市项目的规划策划 [J]. 规划师，2006（11）.

[3] 王志刚. 城市中国 [M]. 北京：人民出版社，2007.

[4] 李德华. 城市规划原理 [M]. 北京：中国建筑工业出版社，2001.

[5] 褚超孚. 城镇住房保障模式研究 [M]. 北京：经济科学出版社，2005.

[6] 汪中求. 细节决定成败 [M]. 北京：新华出版社，2004.

[7] 余柏春. 关于城市美化的几点思考 [J]. 长江建设，2000.5.

[8] 吕明娟. 打造人性化城市公共空间 [J]. 城乡建设，2006（2）.

[9] 黄蔚. 论城市品牌 [J]. 城市发展研究，2005（3）.

[10] 洪生伟. 质量管理 [M]. 北京：中国计量出版社. 2006.

[11] 迈克尔·波特. 国家竞争优势 [M]. 李明轩等译. 上海：中信出版社. 2007.

[12] 刘丽文. 生产与运作管理 [M]. 北京：清华大学出版社. 2006.

[13] 菲利浦·科特勒. 营销管理：分析、计划和控制 [M]. 梅汝和等译. 上海人民出版社，1996.

[14] 孙志刚. 城市功能论 [M]. 北京：经济管理出版社，1998.

[15] G. J. Ashworth, H. Voogd. Selling the city: marketing approaches in public sector urban planning [M]. London: Belhaven Press, 1990.

[16] Porter, M. E. The competitive advantage of nations [M]. London and Basingstoke: Macmillan, 1990.

[17] Saxenian. A. Regional Advantage: Culture and Competition in Silicon Valley and Route 128 [M]. Boston: Harvard University Press, 1994.

案 例

案 例

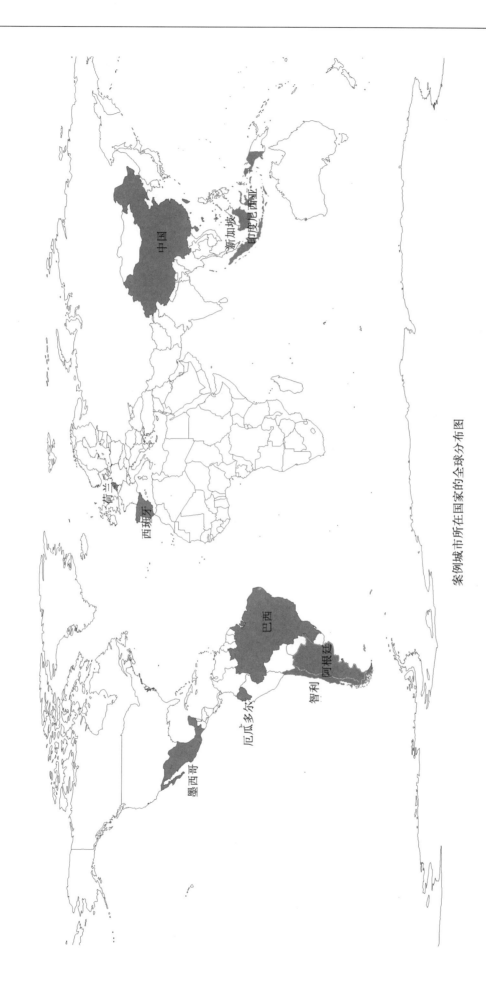

案例城市所在国家的全球分布图

案例

表1 案例城市所在国家的社会经济数据一览表

项目 国家	面积（万km²）	世界排名	人口（万人）	世界排名	人口增长（2000~2005年平均）	密度（人/km²）	GDP（亿美元）	世界排名	GDP增长率（1994~2004年平均）	人均GDP（美元/人，2008年）	世界排名	国民总收入（GNI，美元/人，2009年）	世界排名	农业就业人口比例	工业就业人口比例	服务业就业人口比例	失业率（占劳动力人口比例）	家庭数（万户）	户均人口数（人/户）	人类发展指数（HDI，2007年）	排名
西班牙	50.48	50	4110	28	1.10%	81.4	10400	9	3.40%	31471		32120	41	3.50%	30%	66.40%	11%	1500	2.7	0.955	20
荷兰	4.15	134	1620	61	0.50%	390.1	5790	16	2.40%	45429	13	48460	15	2.70%	22%	74%	6.50%	710	2.3	0.964	7
厄瓜多尔	28.4	71	1362	68	1.10%	47.9	572.5	68	0.40%	3243	97	3970	118	33.10%	42%	25%	9.30%			0.807	77
巴西	851.19	5	18070	5	1.39%	21.2	6040	8	2.20%	6842	78	8040	84	10.10%	38.90%	51%	9.7%(2003)	5160	3.5	0.813	73
阿根廷	276.69	8	3890	32	0.98%	14.1	1530	30	1.10%	6310	57	7550	85	10.50%	35.80%	53.70%	15.6%(2003)	1040	3.6(2007)	0.866	46
墨西哥	197.25	15	104.9	11	1.34%	53.2	6770	14	2.70%	8426	60	8960	78	17%	26.40%	69.50%	2.5%(2004)	2410	4.3	0.854	56
中国	956.09	3	131330	1	0.65%	137.4	19320	3	9.10%	2460	99	3650	125	15.20%	52.90%	31.80%	4.20%(2003)	37810	3.5	0.772	89
智利	75.69	38	1600	59	1.12%	21.1	590	45	4.70%	9698	54	9470	75	6.30%	23%	63%	7.8%(2003)	430	3.7	0.878	45
新加坡	0.064	192	430	114	1.48%	6729.3	1070	42	5.10%	34152	6	37220	33	0%	32.90%	67.10%	5.40%	100	3.5	0.944	27
印度尼西亚	190.44	16	21500(2004)	4	1.18%	128	5160(2008)	18	6.1%(2008)	1824(2008)	120	2050	148	16.3%(2009)	46.8%(2009)	36.9%(2009)	8.14%(2009)			0.734	108

案例城市的社会经济数据一览表

表2

城市	数据	所属国家	人口（万）	占全国比例	人口年增长率	面积（km²）	占全国比例	人口密度（km²）	国民生产总值	占全国比例	生产总值年增长率	人均收入	失业率
毕尔巴鄂	都市圈	西班牙	93	2.31%	生育力指数 0.97	491.6	0.09%	1885.7	77亿美元				
	毕尔巴鄂市		35.9	0.89%		41.3	0.01%	8692					
海牙	海牙地区（无都市圈）	荷兰	92.5	5.67%	-0.24%	82.7	0.19%				3.60%		12.50%
	海牙市区		44	2.69%			0.11%	53.2					11.30%
基多	都市圈	厄瓜多尔	139.8	10.3%	1.6%	324	0.06%	4313.9					
里约热内卢	都市圈	巴西	1087	5.97%	1.14%	5693.5	0.01%						9%
	里约热内卢市		586	3.22%	0.73%	1264.2	0.02%	2381					
罗萨里奥	市域范围	阿根廷	116.2	2.98%	0.36%	488	0.01%	5081					22.80%
	罗萨里奥市		90.8	2.34%	-0.01%	178.7	0.09%	10220					
墨西哥城	城市化区域	墨西哥	1839.7	17.53%		1800	0.04%	12840					
	市区		860.5	8.20%	-0.08%	670	0.07%	2978					
上海	都市圈	中国	1888.5	1.46%		6340.5	2.06%	388.5	13698亿元		9.70%	72534元	4.20%
圣地亚哥	都市圈	智利	606	38.67%	-1.32%	15600	0.53%	6602.4	779亿美元			14389美元	11.40%
	圣地亚哥市		541.4	34.55%	-2.00%	820	0.01%	10330	12.76亿美元			5105美元	
	市中心		25	1.59%		24.2							
新加坡		新加坡	448.39		3.30%	704		6369	2099.9亿新元		7.90%	46832新元	2.80%
雅加达		印度尼西亚	869.96	3.70%		664	0.03%	13100	188.03百万印尼盾		3.98%	1935万印尼盾	

毕尔巴鄂（西班牙）

图1　毕尔巴鄂在西班牙的区位图

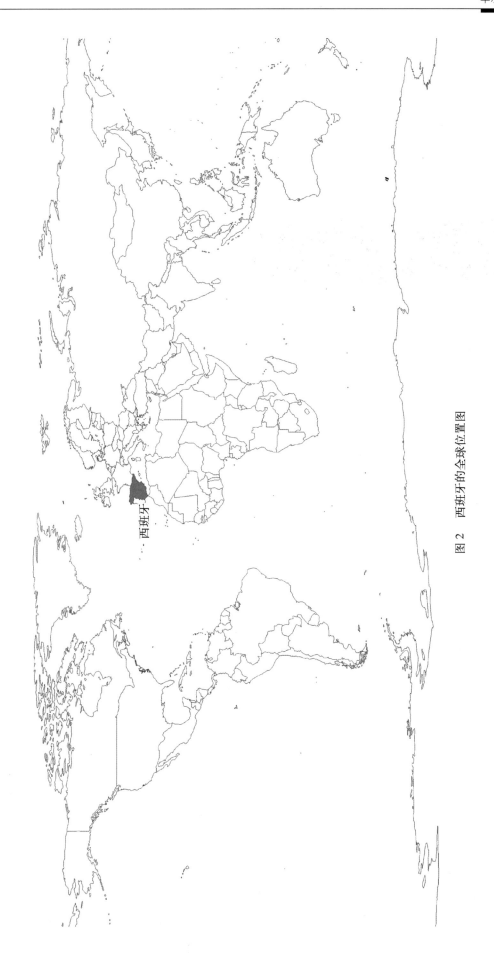

图 2　西班牙的全球位置图

毕尔巴鄂

玛丽莎·卡莫娜 伊斯特班·索托

一、城市数据

(一) 毕尔巴鄂都市圈

面积：491.59km²。

人口：928099 名常住人口（1996 年），927000 名常住人口（2000 年）。

人口增长：生育力指数 0.97。这表明人口增长率明显下降以及老龄化趋势。

在 20 世纪 50～60 年代，毕尔巴鄂的人口数量达到了顶峰，从 20 世纪 70 年代中叶开始下降。到 20 世纪 80 年代常住人口又减少了 30000 人。然而包括市区及郊区的大毕尔巴鄂仍然是整个地区人口最主要的聚居地，集中了毕尔巴鄂总人口的 80%，巴斯克省人口的 44%。

密度：1885.7 人/km²。

巴斯克省的人口总密度是每平方公里 300 名常住人口。

自治区的数量：毕尔巴鄂包括 32 个自治区。

行政区经济活动用地总面积：2405.13hm²。

使用面积：1604.94 hm²。

未使用面积：800.19 hm²。

(二) 毕尔巴鄂市

面积：4130 hm²。

人口：358977 名常住人口（1996 年人口普查）。

总城市化面积：1087 hm²。

住宅数量：136335 套。

密度：135.42 户/hm²。

经济活动用地总面积：143.95 hm²。

使用面积：101.67 hm²。

未使用面积：42.28 hm²。

毕尔巴鄂下属区：Abandoibarra、Amezola、Bolueta、Elorrieta、Elorrieta Canal、Est. abando、Farica Gas、Kadagua、Rekalde、San Ignacio Canal、Zorroza、Zorrozaure、Zorrozgoiti、Gustira。

(三) 都市圈的其他数据

毕尔巴鄂市区人口占总人口数量的 44.4%。

经济活动：

毕尔巴鄂都市圈的国民生产总值 77 亿美元，是巴斯克省的 50%。

第三产业占国民生产总值的 53%，制造业占 32%。

在 1991 年，59.5% 的人口服务于第三产业，31.6% 的人口在制造业，8.3% 的人口在建筑业，0.6% 的人口在农业。

高失业率集中在 Ría 左岸的四个自治区：Sestao 的失业率为 28.2%，Portugalete 是 23.2%，Santurtzi 是 22.5%，Ortuella 是 22.2%。

毕尔巴鄂的局部土地规划预计到 2012 年就业率将增长 18%（新增 52000 个就业岗位），其中 25% 属于传统产业，其余的属于第三产业。

在 1989 年，巴斯克省 26% 的公司已经进行了 R&D（研究与开发）活动。

1999年第一季度技术园区的就业岗位数增长了15%，从5000个到5780个。

1998年增加了8600个新岗位。尽管失业率降低了1.6%，但仍然高达18.7%。

住房：住房需要的统计（DOT，1994）

1991年毕尔巴鄂都市圈的住房总数是334915套。

平均每年住宅建设增长率是6.7%。

在20世纪80年代房屋的所有权中86%是私人所有，10%是租借的，12%是闲置的。

尽管人口数量减少，但对住房的需求将会增加，这是因为新的社会变革产生了新的家庭形式（高离婚率、单亲家庭等等）。

从1994年到2003年，每年有4111个家庭产生，并且占用1190hm^{2}的土地。

独立式住房最多的自治区是Abanto和Ciervana，21%的住房是联排屋。

大多数自治区住宅密度的增加是因为1941年之前由采矿和工业发展所引发的大规模移民。

在1960年Abanto和Ciervana中67.6%的住宅在1941年之前被建造。

毕尔巴鄂市的中央地域是以高比率的空房子为特色：10.8%在Baracaldo，23%在Abantoy Ciervana。

每栋房子人数已经从1960年的4.5人降到1985年的3.5人。

二、城市概况

我们可以断言自1990年以后毕尔巴鄂经历了重要的城市转变，相对于巴塞罗那、塞尔维亚、马德里，以及其他欧洲城市如鹿特丹、柏林、巴黎或里尔来说，毕尔巴鄂这种形态上的转变相当重要。不像其他的西班牙大城市，毕尔巴鄂虽没有经历什么全球事件（如塞尔维亚的四月节、巴塞罗那的奥运会和马德里的欧洲文化首都），却实现了重要的复兴和现代化过程。它拥有独特的位置与一个狭长而扭曲的河口，同时被富含矿产的山所包围，正好进行城市的整合性复兴。在过去很久以繁荣港口和造船活动为特色、以集中采矿业和西班牙产业革命诞生地为形象后，毕尔巴鄂开始着手其形象的根本转变。在几年时间里它已经成功度过了经济萧条、工业衰退和环境消耗，成为一个快速成长型城市。新型、清洁而现代化的工厂开始生产，港口得到再造，主要的基础设施和新的运输系统改善了交通可达性，一个大型的公共卫生计划清洁了河流，建筑物得到改造，更多的公共空间正在建设，各种各样的文化活动得到支持并改变着城市（图3～图6）。

图3　卫星图片
（来源：麻省理工学院网站）

图4　加入拉阿瑞（1893年）与Portugalete设计的
M. A. Palacios当代的埃菲尔挂桥
（来源：Ria，2000）

案 例

图5 Abandoibarra 科技 Barakaldo 都市空中照片
（来源：Ria，2000）

很多人把这一成功归因于城里那些富有远见的机构领导者，他们从20世纪70年代起提出了多个城市发展设想与规划，并利用危机带来的机会和政治与制度改变带来的机会，使毕尔巴鄂调整区域发展目标以找到在欧盟的地位，并为引导发展设立了两个城市公司。

在20世纪90年代后期城市已经变得现代化，土地利用得到调整，港口得到搬迁，破败的工业和旧的建筑物得到修复使用，而且在Ría河周围的大片地区得到重建。通过复杂的财务安排，许多大规模开发项目开始实施，其连锁反应是本地企业开始获得收益，可达性得到改善，并吸引了国内和国际的投资者。同时，市民对未来充满信心，世界各地的游客则蜂拥而至（图7～图9）。

毕尔巴鄂所在的Ría河位于伊比利亚半岛的北部腹地，从毕尔巴鄂到坎塔布里安海有15km长，贯穿了两岸复兴中的城市。那里有一个由Lucero角和Galea角形成宽度达5800英尺的河口，使毕尔巴鄂利用其与欧盟各国邻近的优势，在西班牙贸易中起到核心角色。古根海姆博物馆是最早沿Ría河建造的战略性项目之一，项目所在的用地从前是堆放Barakaldos的冶金合成物，这个钛制建筑物已经成为城市转变的动力，并且引发了许多有关文化能如何提高城市再生和城市经济发展的辩论。这种新发展不同于1960年

图6 在河口和毕尔巴鄂港的航空照片
（来源：双圆巴斯克）

图7 石版画，作者未知

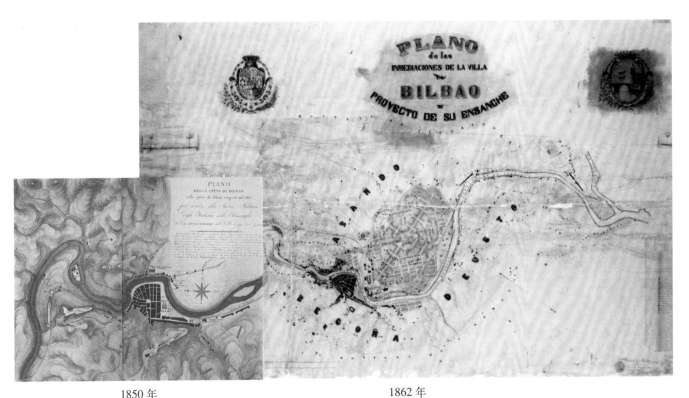

1850年   1862年

图8   1850年和1862年的城市规划图
（来源：Ria，2000）

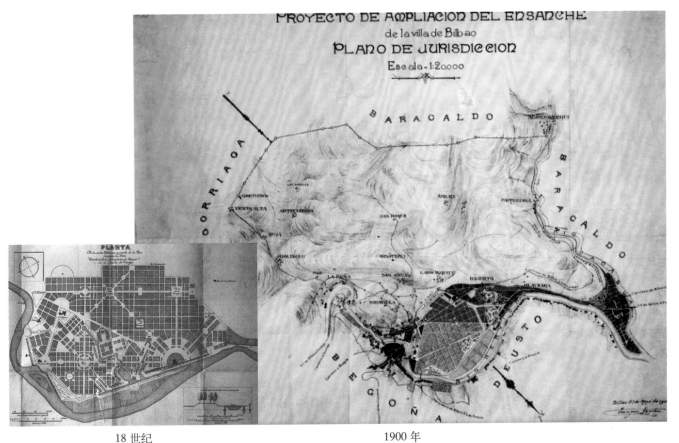

18世纪   1900年

图9   18世纪和1900年的城市规划图

代那种以空气污染和低失业率为特征的旧城发展。从 1977 年到 1988 年间，Vizcaya 省的 41.1 万名工人受到工业危机及其后产业结构调整的影响，随之而来的是更多岗位消失，很多移居而来的人不得不返回其出生地。

毕尔巴鄂区域以其人口高密度和住宅高密度为特征，直到 1960 年毕尔巴鄂市的主要住宅类型都是像香港一样的集中式高层建筑，又以 Portugalete 和 San Vicente de Barakaldo 为甚。20 世纪 40 年代的战争造成了住宅建设的停滞，这导致了 20 世纪 50～60 年代在 Barakaldo、Portugalete 和 Sestao 的建设热潮。

毕尔巴鄂的失业率高居巴斯克省第二，其原因既可能是工业结构转变的结果，也可能是第三产业没法吸纳更多的劳动力。在 Ría 河左岸的高失业率伴随的暴力、吸毒和社会问题是毕尔巴鄂市政当局必须面对的最主要问题之一，而其对策是促进文化发展，推动社会与经济变革，发展旅游业，以及发展洁净的工业和服务，所有这些都为左岸那些目前依赖非正规经济、当地小企业以及其他方式生存的人口创造了新的机遇。

三、城市历史

最早的巴斯克居民点历史出自远古文化：Vascones、Caristios、Bardulos、Autrigones 和 Cantabros。毕尔巴鄂在 1300 年开始出现独立住宅，中世纪国王与贵族之间的冲突影响到领地的划分。19 世纪初，像其他欧洲区域一样，毕尔巴鄂以中央集权管理、新闻媒体形成、商会发展、土木工程师出现以及大型工业设备出现为特色。毕尔巴鄂市也受到英国工业革命、法国政治革命思想，资产阶级初期的发展和民族主义及其他资产阶级意识的影响（图 10）。

有两个战略项目始于 19 世纪，包括：La Paz 港以及 Ensanche（城市的延伸）。La Paz 港作为自由港建设，在那里可以独立于企业协会与贸易协会之外设立新的企业。1876 年，毕尔巴鄂的扩展计划（Plan de Ensanche）开始实施，这个计划主要是开发河右岸的 Abando 基地（包括 Alzola、Achucarro 和 Hoffmeyer）。与马德里、巴塞罗那和圣塞巴斯蒂安相类似，毕尔巴鄂的扩展计划由一个可扩充的斜线网格构成，但没有考虑到地理上的决定因素。由于该地的形态和港口活动的位置关系，这一扩展计划并未触及到河岸本身，与当时其他的城市干预项目一样，港口区域与城市结构存在着明显的分界。

采矿业和冶金业对劳动力的需求吸引了来自整个西班牙北部的移民，虽然有花园城市、廉价房等新的手段，移民的住房仍然以低标准和高密度为特色，城市的生长则形成了延伸到海边及河岸两侧的村庄，整个城市带人口达到 100 万左右，毕尔巴鄂市则为 33 万人，城市形成了单中心的形态。虽然没有预先的安排，但自然的分隔已经产生：矿区、高炉、船坞位于左岸，污染少阳光多的右岸则成为资产阶级的住所。

毕尔巴鄂最大的城市变革是在 1968 年出现，因一系列相关的运作而产生，是一个产业运作的副产

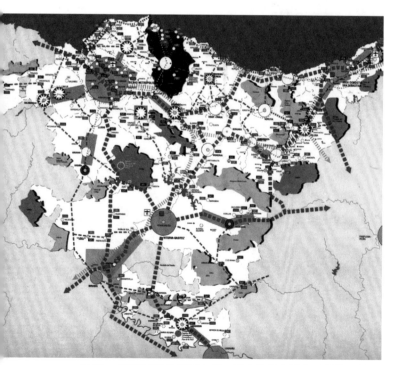

图 10　区域运输的指导方针，层次结构和城市系统的连接
（来源：DOT）

品。事实上当Somorrostro的石油精炼厂设立在河口东部海岸上的时候，港口及城市的巨大变化才成为可能。在这个市政项目的公开投标中，中标者引入了建设Lucero角大堤的想法，使得大型油轮泊位的建设成为可能。这一决策在30年后带给毕尔巴鄂重振经济和再利用废弃的工业区以巨大的机遇，新的项目如Abandoibarra、古根海姆博物馆、Euskalduna的建设才变得可能。

20世纪毕尔巴鄂的发展以矿产开采和工业发展为特征，前半个世纪工业的发展是以钢铁工业为基础，主要依靠河左岸的铁矿。伴随着地表氧化铁矿床的开采耗尽和国际市场对铁需求的急剧减少，采矿活动从19世纪末开始减少，其间伴随着两次世界大战、西班牙内战以及逐渐加剧的劳动力冲突。由于地下采矿业的高额成本和钢铁工业昂贵的处理技术要求，国家没法解决这种危机。对技术和投资的要求，使小型矿主消失，代之以大型的私营联合体。

20世纪的前半段，产生了一些大型的工业企业，例如Viscaya高炉公司、船运公司、S.E.C、Babcock和Wilcox等。直到20世纪70年代中期，这些工业的发展创造了大量的财富和充足的就业岗位，这也是导致空气、土壤和水严重污染的原因。然而生产结构对大型传统企业的依赖以及这些企业与易受危机冲击的部门诸如采矿业、钢铁产业和航运业的联系，使整个城市经济在20世纪70、80年代的经济危机中遭遇了严重的创伤（图11）。

工业化的过程中形成了一个网络，网络中中小企业依赖着大公司而发展，并在遭遇危机的时候随着这些大公司一起倒闭。在超过15年时间里，巴斯克省的经济恶化和经济合作与发展组织（OECD）的各国有关，并在这些区域与欧盟的整合中遭遇很多困难。从1987年起，毕尔巴鄂的经济走上了复苏的道路，现代化的步伐开始，市民信心也随之恢复。许多大公司进行了工业再造、现代化和自动化，新的协同作用开始发挥作用，以工业化为导向的服务业、金融和清洁的高科技工业开始繁荣发展。Vizcaya高炉公司的一个主要技术转化项目和新港区设施的建设项目吸引了来自欧盟各国大规模投资的追逐（图12）。

图11 城市传统中心航空照
（来源：Ria，2000）

（一）规划历史

在佛朗哥专政时期市政府的权力因更大尺度的全球背景的规划（Plan Comarcal）而受到约束（1943年）。1943年设立了大毕尔巴鄂公司，并根据CIAM（国际建筑师协会）的理论为毕尔巴鄂的21个自治区制定了都市圈总体规划。然而这个跨自治区的规划没有达到可供实施的程度，原因是它缺乏城市规划文化，以及存在资金上的困难。

城市的发展伴随着社会住宅建设的步伐，得到了大企业、私人公司和住房部门的支持。社会住房项目通常都是高密度建设，靠近现有的高速路并位于那些为了住宅发展而安排的特别区域内，但缺少配套的道路系统和城市服务体系。所有的这些新住宅区依据"部分人口服务计划"而建造。事实上，在城市一级缺乏住宅土地使用导引，而住宅建设不得不依据城市总体规划纲要来指导。

案 例

图12　新的技术和工业楼宇
（来源：Ria，2000）

1956年前，城市规划被若干无法确保整体性与空间一致性的几个部门所控制，直到那时立法都是由城市扩展法案控制（1892年）、公共卫生和室内改善方案（1895年）以及房屋土地管理机构（1945年）和地方政权条款（1955年）制定完成。1956年国家土地使用和城市调整规划批准通过对先前法案缺陷的修改。这个法案被视为"不足以应对和解决城市秩序，以满足未来西班牙人口增长与科技发展的需要"。同时，这种跨地域的城市发展纲要无法控制越来越严重的密度问题。

地方级立法和城市管理系统的匮乏促成了首个毕尔巴鄂都市圈总体规划在1964年的制订，这个规划指导了工业和基础设施的发展，尤其是交通业和通信业，不过这个规划仍然比例过大（1∶10000）。在这个比例下土地使用区划不能反映土地和建设程序的差异，缺乏对地方道路、开放空间及服务系统的界定。随着城市的继续成长，超出了城市发展条例的规定，城市结构呈现出混乱和整体上的不足。在19世纪60、70年代，一种新的城市意识开始产生，邻里关系运动的出现，环境污染问题受到重视，在邻里的层面上寻求解决环境污染的办法得到推崇。佛朗哥的死亡（1975年）、新的西班牙宪法通过（1978年）、为巴斯克省自治而制定的地方性民主化法令（1979年）、第一个巴斯克事务委员会和巴斯克议会的产生，这些历史性的里程碑，它使巴斯克省的复兴变得很有可能，西班牙加入欧洲共同体（1992年）也影响了整个复兴的进程。

随着一部新的土地使用和城市管理法规（1978年）的实施，以及城市权力向巴斯克省的转化，上述状况开始改变。新的民主宪法制订后，西班牙开始着手从中央向地方政府的权力分散，确认了不同层次城市政府拥有在公共投资及司法管辖上的权限。毕尔巴鄂的总体规划（1998年）和实施规定（DOT）则是指导现阶段城市变化的新手段。

政府为降低过高的密度而制定了operaciones de esponjamiento de tejidos计划，而疏解工业拥挤的计划则在区域范围内推动了新的设施建设及高附加值的产业活动。

1975年的城市民主化带来了历史性的变化，但这一复兴的过程由于工业模式的失败和1983年发生的五百年一遇的大雨所造成的经济萧条而推迟。这场大灾难或许在毕尔巴鄂历史上的这一时刻成为城市复兴的导火线，多种因素的聚集使这个具有重要意义的城市复兴规划的发展条件成熟，形成了连贯的地域模式。

如果奇迹存在的话，自然有它存在的原因。很明显，20世纪80年代的城市化——仅仅只是地方政府的手段——包括不协调的干预和对城市结构的局部改善，都无法为城市复兴奠定明确的基础。为了发挥应有的作用，不得不制订基础设施发展的毕尔巴鄂Ría2000计划，仿佛转瞬间完成了新的港口、新的机场、物流规划和城市公共卫生、地铁、道路系统。

四、全球化，机遇和挑战

对于巴斯克人来说，经济国际化已经变成一个基本的议题，巴斯克地区把重心集中在科技创新、基础设施以及新的管理和营销程序上。截至1989年，毕尔巴鄂超过四分之一的企业实施了研发活动，各种技术中心的研究领域涉及机器人技术、新材料、电子、化学和环境科学等。在Zamudios科技园中，既有高科技的大企业，也有许多因为规模小而需要国家扶持的R&D中介服务企业。毕尔巴鄂都市圈的34个自治区都为新的服务业、旅游业、高科技工业提供了新的工作场所，并特别关注信息和通信技术的普及（图13）。

这一城市群通过加强它作为城市增长极来提升自身的地位，并开拓出更高级的功能，例如通过提供更多对企业的支持服务来提高技术专门化程度。这可以帮助毕尔巴鄂将活动辐射至周边邻近的城市，并加强巴斯克地区与欧洲最活跃地区之间的联系轴与通道。为了实现这种特殊性，非常有必要改善基础设施服务，加强与其他欧洲地区在技术和贸易领域的交流。

这个地区可以成为欧洲范围内的一个战略性节点，因为：

（1）它在欧洲地域内的中心位置；
（2）它位于从北欧到Andalucia、巴黎、Aquitane、巴斯克地区和马德里的走廊上；
（3）巴斯克地区位于大西洋地带，靠近欧洲最活跃的地区。

在毕尔巴鄂附近有几个重要的城市，整体上它们形成一个欧洲中心点，成为环大西洋地区与欧洲最活跃地区之间的连接点。

公共和私营部门迅速认识到毕尔巴鄂面临的机遇，两个组织——毕尔巴鄂Ría2000、毕尔巴鄂Metropolis 30分别成立，以充分利用其区位条件，并对宏观经济现实带来的机遇有快速的反应。

毕尔巴鄂Ría2000是一个匿名的公共基金组织，

图13　ITP发动机厂，在Zamudio技术公园中
（来源：Ria, 2000）

成立于1992年。这笔资金由国家各行政部门通过其拥有的企业（SEPES土地公共事业、西班牙开发部、西班牙铁路RENFE、西班牙工业部、毕尔巴鄂自治港），以及巴斯克地区行政部门（Foral各县地方政府以及Ría的各自治区）共同组成。

毕尔巴鄂Ría2000正在发展的项目有：铁路大道，在Amezola的公共投资，在Abandoibarra的公共投资（Esukalduna和古根海姆博物馆），在Barakaldo的公共投资。

毕尔巴鄂Metropoli 30是一个公共/私营合作的机构，它试图促成整合，以及为毕尔巴鄂的复兴而规划和培育未来的项目。

当前的繁荣创造了新的就业类型，老工厂被改造成高档艺术中心和文化场所，Samudio科技园吸引了60家高技术企业，并创造了2500个新的就业岗位。

毕尔巴鄂的改变促进了文化、旅游和服务业的发展。1998年的游客数量达到1997年的两倍并保持了持续的增长。艺术设计和建筑设计已经成为促进城市发展、更新和创造就业机会的新要素。世界著名建筑大师弗兰克·盖里（Frank Gehri）设计的钛宫殿（译者注：即古根海姆博物馆）、卡拉特拉瓦（Calatrava）设计的步行桥、诺曼·福斯特（Norman Foster）设计的地铁连通设施、Soriano和Palacios设计的国会大厦，都是吸引各种大型活动的标志性建筑。

（一）使城市大规模运作获得利益的战略和手段

在毕尔巴鄂和其他西班牙城市，规划师们了解各级政府之间与各政府部门之间的协调对于形成地方整合发展的重要性，因此极力吸引公共和私有投资来推动毕尔巴鄂必需的城市改造（图14、图15）。

毕尔巴鄂Ría2000代表了一种模式，即采取怎样的战略与手段来实现多种目标，其管理委员会由20名成员组成，代表了他们各自所在的部门，主席由毕尔巴鄂市市长Iñaki Askuna担任，副主席是国家基础设施和规划部的部长Victor Morlan Gracia。该机构的使命是复兴毕尔巴鄂市区内衰落的区域和废弃的工业用地。为实现这些目标，该机构协调和执行了诸如城市建设、交通和环境方面的公共建设项目。这些项目的开发采用了全球招标的方式，所有的公共管理部门都参与其中。

各自治市认为该机构必须使土地能够适应新的用途，并为此将资产（包括港口和RENFE铁路）移交给毕尔巴鄂Ría2000，以开始初步的准备工作，包括进行土地配套，然后将土地出售给私营开发商根据规划修建住宅、商业或办公建筑。这些营运获得的盈利——通常因为项目位于中心地点而数量可观——则投入建设新的项目，使其他面临衰落的地方得以复兴。

这个机构的起源要追溯到1987年，毕尔巴鄂市政当局制定的第一份城市调整规划指出城市发展的契机在Abandoibarra和Ametzola，那些河畔的国有土地。机构展示了它的能力，没有占用公共投资就实现了财务平衡，并得到了欧盟的资助。创造利润的策略就是改造旧区和建设现代化的铁路系统。Variante Sur、毕尔巴鄂La Vieja和Barakaldo计划就是这一操作模式的实例。所有都市化进程都按照和谐发展的目标，在对未来城市图景的清晰了解中完成。这些项目的城市设计质量更增加了该区域的价值，甚至超过其土地因区位而具有的价值。

在毕尔巴鄂，区域规划和产业规划综合研究了国土、地形、景观，并有人口及统计数据的支撑，对实现城市发展目标发挥了必需的作用。同时，针对土地开发规划、人口与空间规划、区划、绿地与公共空间规划、保护区约束条例、水资源及混凝土容量的研究成为城市预测的依据，并描绘了城市的未来景象。根据巴斯克城市传统，建筑表达和城市设计是这个分析过程与政策制定的一部分，城市项目则应该向市民呈现未来城市的形象。

图14　卡拉特拉瓦设计的桥

图15　都市圈用地规划确定的两条滨水高品质公共空间
（来源：都市圈用地规划）

(二)用地调整的原则

规划问题的出现是由于缺乏明确的用地及机制上的指导框架,需要通过用地调整原则在公众心目中为城市建立共同的未来形象,主要元素包括:非连续性发展过程、城市沿着Ría河断断续续地发展并形成线形集聚;城市因河流的分隔而需要更多的桥梁;边界要素的重要性;地下空间系统等。

规划共分为三个层次,第一层次是针对整个巴斯克区域(包括LOT和DOT),中间层次是毕尔巴鄂的功能形区域、圣塞巴斯蒂安以及维多利亚地区,第三个层次是形成毕尔巴鄂都市区的32个自治市。

在整个巴斯克区域这个层次中,用地调整原则(DOT)的主要挑战是强化用地模式,巩固城市系统对经济、社会和文化创新在欧洲最具活力地区的吸引力。人们相信,规划能在整个巴斯克地区带来均衡的发展机会。为了达到这个目标,需要为城市系统制定不同层次的战略,使最主要的三个城市——相互距离很近的毕尔巴鄂市、Donostia-San Sebastian(圣塞巴斯蒂安)和Vitoria(维多利亚)-Gasteiz——能以一种平衡的方式推动整个巴斯克地区的复兴,并通过一个连贯的高速公路系统推动经济发展和解决现有的不均衡问题,这一尺度上的规划意在加强次一级的城市与毕尔巴鄂及维多利亚的联系(图16、图17)。

(三)毕尔巴鄂大都市的战略远景

巴斯克经济的复苏要靠毕尔巴鄂市实力提升的带动,并与周边城市更好地整合。毕尔巴鄂是西班牙全国仅次于马德里的重要城市,许多重要的城市都在它的区域影响下(如桑坦德、圣塞巴斯蒂安、维多利亚、潘普洛纳和洛格罗诺)。最新的人口普查显示毕尔巴鄂市中心的人口在减少,作为巴斯克多核城镇体系的重要节点,毕尔巴鄂市的强大要靠以下几点为基础:

完成基本的基础设施(机场、港口、铁路系统、道路系统等)以建立一个与国际国内经济发展相连接的节点与轴线。

注:之后,每个地区由不同的团队进行了开发,但都是基于同样的原则与干预措施

图16 毕尔巴鄂都市区用地规划中所载的一些初步想法
(来源:巴斯克政府DOT)

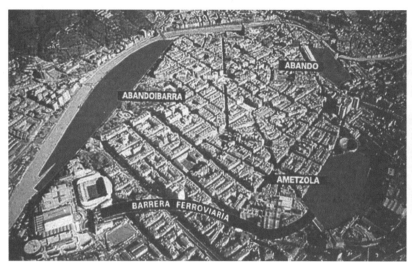

图 17　城市中主要轨道交通干线及站点
（来源：DOT）

图 18　人造岛屿

最大程度上强化毕尔巴鄂，特别是 Ría 的区位对于全国服务与活动的重要性，作为巴斯克地区最中心的节点，这必然会影响到土地利用的选择。

制定一个计划来改善城市环境，特别是实现必要的经济变革。通过对 Ría 附近空间的改造，它们不再是城市的边缘，而会成为新用途发展的轴线，以及各个市区城市组织的中心（图17）。

毕尔巴鄂都市圈的复兴战略规划是在 1989 年应巴斯克政府和 Bizkaia 县议会的请求开始的。人们相信，在都市区的各种不同经济与社会力量之间，无论强势还是弱势，相互包容共存是可能的事情，同时，可以建立城市的未来形象并制定实现它的战略。这一形象预示着毕尔巴鄂将成为开放、复合、整体、现代、创造性的、社会的及文化的都市。这个战略规划还强调了对人力资源的投资：一个现代、灵活而创新的教育系统，一所与大城市经济结构紧密相关的大学，一项有关人力资源事业的战略构想，以及具有领导能力的公共事务管理（图18）。

局部用地规划提供了一个使得不同的复兴干预活动合理化、和谐化，且通用而有效的框架，力求纠正现存的无数不均衡状态，其全球性的目标如下：

● 整合 Ría 两岸，并最大限度地利用河滨建设高质量城市空间；

● 减轻当前城区高密度和过度拥挤的问题，特别是在 Ría 左岸和 Basauri 地区；

● 促进设立经济活动区，在这个区域内工业和服务业混合设置，并互相促进；

● 更新被逐渐衰落的住宅和工业区，通过创造新的经济活动（服务业、制造业），为高密度旧城区的结构调整提供新的都市型式、设施和开放空间，以改善邻里空间，促进住宅和服务业的共存；

● 复原古老或被遗弃的矿区；

● 优化机场和通信设施周边的用地并用于新的发展，并建设大学与机场和科技园区之间的发展轴线；

● 控制 Asua 山谷地区自发形成的工业发展；

● 引导在 Getxo 和 Plentzia 之间沿海地区的发展，创造优质的聚居环境；

● 创造一个贯穿全市区的公园系统，包括河岸两侧恢复的绿地（图19、图20）。

五、城市战略

（一）运输系统

联系系统已被用来优化外部和内部的连接。与大西洋的连接被设想为欧洲一条重要的轴线，并且巴斯

图 19　初步构想　　　　图 20　局部鸟瞰　　　　图 21　1980 年城市的中心港口和船厂
（来源：Ria, 2000）

克地区可以在这个轴线上处于一个关键角色。内部流通在多核的巴斯克地区的城镇间形成高效率的道路网络，并通过连续的细小网络与外界相连。用地调整原则（DOT）建立了内部形态系统，并旨在加强城市社区与郊区的全球联系。以下是内部形态系统设计的主要原则：

（1）巴斯克系统与欧洲城市的连接。有些交通联系是高速度和高容量的：布尔戈斯－马德里，Cornisa Cantabrica，通过 Logrono 和潘普洛纳的 Ebro 通道，以及 Donostia-Baiona 通向 Burdeos 和巴黎的走廊。

（2）巴斯克多核城镇体系与高容量道路系统和高速铁路系统的相互联系。

（3）不同功能区域的主要城市与各中等城市的强大联系，用以实现生产体系的整合，以及更好的社会与文化联系。

（4）发展主要城区的高容量公共交通系统，尤其是毕尔巴鄂地铁。改善毕尔巴鄂都市圈的可达性，特别是从 Donostia——圣塞巴斯蒂安方向。通过优化巴斯克地区的机场体系，均匀地覆盖毕尔巴鄂、维多利亚和圣塞巴斯蒂安、比亚里茨和潘普洛纳等城市。

（5）通过优化基础设施和改善对欧洲主要口岸的联系实现毕尔巴鄂港的现代化（图 21）。

（6）最后，DOT 提出大量改善通信和能源基础设施的计划，这被认为是具有很高的战略价值。

（二）基础设施的规划和整合

整合基础设施的规划使变更土地利用成为可能，高速公路网络能够适应新的需求，供水和排水系统得到全面升级，地铁系统沿着一条旧的铁路线得到翻新，还在左岸建设了新的地铁系统。在 Abra 外围的港口建设、Sondika 机场、Zamudio 科技园的新建，工业土地的迁离以及铁路系统向南引申，所有这一切都促进了 Ría 都市空间的发展（图 22、图 23）。

在毕尔巴鄂 Ría2000 介入铁路系统后，雄心勃勃的 Variante Sur 项目得到实施，这个计划重新调整了城市的铁路系统来连接左岸与毕尔巴鄂中心和地铁系统。这个项目促成了四个新火车站（Zabalburu, Ametzola, Autonomía, San Mamés）的建设，以及 Abando 站与 Olabeaga 站的更新。毕尔巴鄂 Ría2000 已经在 Pena 社区建立了一个新车站，下一个将位于 Miribilla 的新城区，并将在西班牙制造部和铁路公司的合作下，在 Santurtzi 建成一个新的城市间火车站（Cercanías RENFE）。

旧地面铁路变成地铁及城市间火车线路（líneas de cercanías）的建成意味着新 Avenida del Ferrocarril 的出现，它将 Ensanche 与两个重要社区连接在一起，消除了路面的城市障碍。San Mamés 站中心换乘枢纽的建成将地铁、城市间火车、公共汽车和电车系统连接在一起。

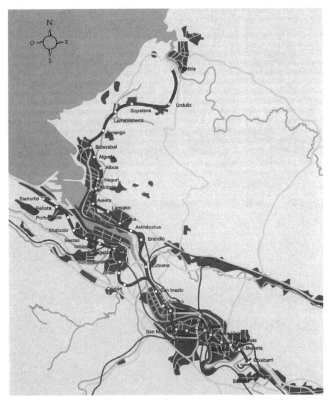

图 22　城市主要的铁路线和主要车站
（来源：DOT）

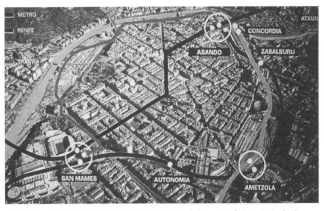

图 23　毕尔巴鄂的河口部分区域计划，图中圆圈处为确立的"发展机会区"
（来源：毕尔巴鄂都市处）

（三）Ametzola

这个开发项目的建成将过去是三个货运车站的老中心地区改建为拥有 900 套新住宅和 36000m² 城市公园的新中心。

（四）旧毕尔巴鄂（Bilbao La Vieja）

毕尔巴鄂 Ría2000 致力于复兴那些当前已陷入破败和边缘化过程中的历史中心地区，由于周围其他区域的改造，历史中心区开始受益于土地等价值的提高。规划和实施的几个新项目包括带有地下停车场的 Corazón de María 广场，改造成公共步行空间的老码头，以及 Catalojas 桥附近的连接区改建成的广场。

（五）Galindo – Barakaldo

根据将大规模项目创造的收益用于改造旧区的原则，Barakaldo 区策划了一个占地达 50hm² 的城市运营计划，如同毕尔巴鄂城中的阿班多巴拉（Abandoibarra）一样。这个庞大的计划将城市结构延伸到 Nervion 河与 Galindo 河的河岸，包括一个市政环路的道路系统与各种各样公共建筑的建设，例如 Desierto 广场、Barakaldo 足球场以及新体育中心。

毕尔巴鄂展览中心也在 Barakaldo 落成，它是全国性的标志项目，由两个巴斯克的工程设计事务所设计建造，占地 40 万 m²，包括了展览馆、停车场和办公区，是激发 Barakaldo 地区复兴的重要动力。

（六）桥梁

桥梁是毕尔巴鄂历史上一个重要元素和象征，虽然它们从未实现 Ría 两岸真正意义上的整合。1893 年 Eiffel（埃菲尔铁塔的设计者）的一个学生 Martin Alberto Palacios 设计、建造了连接 Portugalete 与 Las 体育场的悬索桥。1997 年作为 Euskalduna 一部分，Javier Manterola 设计了一座辐盖形桥梁，卡拉特拉瓦设计了一座白色人行桥（Zubi Zuri）。弗兰克·盖里设计的古根海姆博物馆则集成了 Salve 桥和 Fernandez Deusto 大学桥。

图24 旧的吊桥

图25 白桥，由卡拉特拉瓦设计

图26 径向桥，由Manterola设计

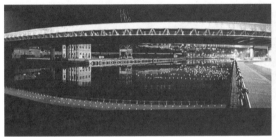
图27 由Manterola设计的径向桥

图28 毕尔巴鄂地铁
（来源：埃斯特万·罗德里格斯）

图29 毕尔巴鄂地铁设计
（来源：埃斯特万·罗德里格斯）

人行桥Pasarela Pedro Arrupe是由建筑师Jose Antonio Fernandez Ordonez始建，他的儿子Lorenzo和工程师Francisco Millanes继续完成的。它由钢木复合建造，内层材料为重蚁木。这种视觉上的对比——冷酷的外表和温暖的内部——成为它最受赞赏的特征之一（图24～图27）。

（七）地下铁路

毕尔巴鄂地铁从1997年开始正式运营，富有吸引力的入口和车站由诺曼·福斯特（Norman Foster）设计，不仅提高了城市的可达性，也提高了城市生产力，改善了城市形象（图28～图30）。

（八）新的港口

在Abra外部建成了一个具有现代化设施的新港口，使位于现在城市中心的旧码头区用地获得了重新开发的机会。这些新设施与其他运输系统和加工企业方便地连接，使得港口成为城市更新和发展的重要动力之一。

（九）飞机场

Sondika机场（建筑师卡拉特拉瓦设计）提供了城市与大西洋及延伸地区、国际和国内航线的空中连接。一个新的现代化机场航站楼已经建成，1999年接待了400万名乘客，预计2025年将接待乘客1000万（图31～图33）。

六、城市大规模开发项目

（一）阿班多巴拉的更新

Ría区域最大规模的更新发生在阿班多巴拉，它是一个在毕尔巴鄂市中心占地346000m²的区域，那里1/3是公共开敞空间（115000m²）。它是城市最典型的项目，从建在Evaristo Churruca老码头上的Paseo de

案例

图30　毕尔巴鄂地铁出入口设计
（来源：埃斯特万·罗德里格斯）

图31　圣地亚哥·卡拉特拉瓦的机场设计
（来源：GEO，德国国家地理，1998）

图32　圣地亚哥·卡拉特拉瓦的机场设计
（来源：GEO，德国国家地理，1998）

图33　圣地亚哥·卡拉特拉瓦的机场设计
（来源：地理，1998）

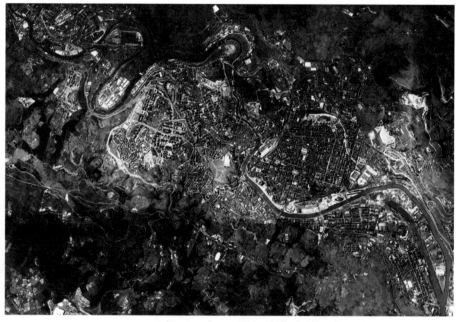

图34　Abandoibarra与Barakaldo都市空中照片
（来源：Ria，2000）

Ribera河滨大道开始，毕尔巴鄂市民和游客能徜徉在过去用于工业生产的用地上，包括造船厂、一个集装箱库区、两个火车站及一个码头。当年，随着这些设施的关闭，这一地区迅速衰落。整体区域的复兴由毕尔巴鄂Ría2000来主持。在古根海姆博物馆周围的区域中又完成了两个文化设施：建筑师Dolores Palacio和Federico Soriano设计的Esukalduna宫以及建筑师Juan F. Paz的SENER事务所设计的Maritime（海洋）博物馆。

根据建筑师西萨·佩里（Cesar Pelli）、Diana Balmori、Eugenio Aguinaga组成的一个小组提交的局部规划，整修工作于1998年开始，计划必须以两栋主要建筑——古根海姆博物馆和Euskaldina宫为主。阿班多巴拉大道的第一阶段已经在Euskalduna宫和Pedro Arrupe、Paseo de Ribera步行桥之间对公众开放，一个儿童游乐区和计算机控制的喷泉也已完成。这个公共步行空间栽满了树，旁边还有一条自行车道和三个电车车站。大道由台阶连接到Deusto桥，还开辟了Lehendakari Leizaola街与Deusto桥之间互通的机动车道。阿班多巴拉大道的第二个阶段正在建设，它将直接连接到Uribitarte。

整个规划项目共计花费1.8亿美元，26%由毕尔巴鄂Ría2000资助（通过将土地出让给私人企业获得），45%用于建造古根海姆博物馆，29%用于建造音乐厅，均由省和地区出资。古根海姆博物馆建筑面积为25000m²。Euskalduna音乐厅有25000m²，包括 2200 座的演奏大厅、三个小厅、八间训练房，以及七间会议及新闻发布室、咖啡厅、餐厅和商店。由 ING-Sonae 财团出资、建筑师 Robert Sterm 设计的 Zubiarte 商业中心有 25000m² 的商店和娱乐空间。建筑面积 14000m² 的豪华喜来登（Sheraton）饭店由 Ricardo 和 Victor Legorreta 设计，一座不锈钢制、连接 Pasarela 大学与 Deusto 大学的阿班多巴拉桥由建筑师 Jose Antonio Fernandez 设计，建筑师 Javier Lopez Chollet 设计的 Ribera 公园是一个长达 3km 的雕塑公园。Maritime 海洋博物馆在 2003 年投入使用，容量 250 人的电车系统也已完成（图 34～图 46）。

阿班多巴拉的系列改造项目尚未全部完成，已与其他沿河地区的新发展一起将 Ría 变成了毕尔巴鄂市的中心，西萨·佩里设计的办公楼是此地区规划的结

图 35　有所改善的新车站，包括一条新的街道

图 36　地下的火车线路
（来源：Ria, 2000）

图 37　Euskalduna 宫，由帕拉西奥多洛雷斯和费德里科·索里亚诺设计
（来源：E. Rodriguez）

图 38　Amelola 大街

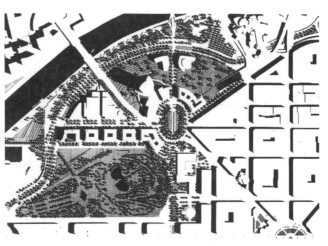

图 39　旧工业区改造方案
（来源：Gobierno Vasco）

案 例

图40　1990年跨毕尔巴鄂一个开放的铁路通道

图41　在建的 Euskalduna 宫
（来源：埃斯特罗德里格斯·索托）

图42　弗兰克·盖里设计的古根海姆博物馆
（来源：Glenda Kapstein）

图43　弗兰克·盖里设计的古根海姆博物馆
（来源：Glenda Kapstein）

图44　毕尔巴鄂港河口的航空照片

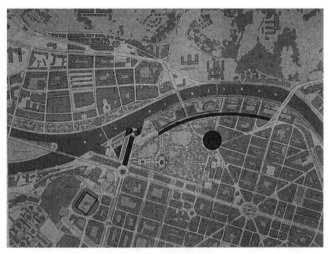

图 45 塞萨尔计划
（来源：BR，2000）

图 46 从古根海姆博物馆到 Euskalduna 宫
（来源：Ria，2000）

束之作，已成为城市金融中心所在地乃至欧洲大西洋地区最杰出的建筑之一。阿班多巴拉内还有一幢巴斯克大学神父寓所办公楼，800 套新住宅和 81000m² 的高标准办公楼，以及 Deusto 大学的新图书馆。

好几栋建筑已经获得国际知名度并且获得了奖励，如诺曼·福斯特设计的地铁车站，弗兰克·盖里设计的古根海姆博物馆，Eskalduna 宫被授予最佳国会大厦设计奖。毕尔巴鄂 Ría2000 因为其对 Ría 沿岸实施的名为"产业边缘、城市边缘"的复兴而在威尼斯双年展上获得 CITTA D'ACQUA 奖。

七、结论

毕尔巴鄂，曾经的工业城市快速变成为文化与多功能的城市，其原因在于为数众多的机构与个人预见到城市面临的重要结构与功能转变。在 20 世纪 60 年代末深水港建成以及 1992 年西班牙加入欧盟后，被遗弃的工业场地成为非营利机构进行创造性再开发活动的场所（特别是毕尔巴鄂 Ría2000），其目标是改造城市中心的废弃地并使它们重新获得生机，使城市河滨两侧获得均衡的发展。随后政府制定了新的城市景观规划，吸引了广泛的投资者和知名设计师，其结果是毕尔巴鄂的更新改造成为整个欧洲最为出色的案例。

阿巴多巴拉的 Ametzola 是其中的首个重点项目，它为后续的项目在目标、策略、运营管理、合作方式等方面提供了良好的借鉴。该项目集成了建筑与基础设施的开发，古根海姆博物馆、Euskalduna 宫酒店、海洋博物馆，以及连接 Nevro 河两岸样式各异的桥梁，都使毕尔巴鄂成为一个值得一游的文化城市。

参考书目

[1] Teresa Segura（集市）与 Esteban Rodriguez（SENER）的采访录（1999）。
[2] Directrices de Ordenación Terriorial (DOT) de la Comunidad Autonoma del País Vasco[Z], 1994.
[3] GEO, especial. Bilbao en Vanguardia[J], 1998.
[4] El Mundo. El Libro de Oro de Bilbao. Siete Siglos de Imagenes[Z]. El Mundo, Homepage Bilbao Metropolitano, 1999.
[5] Directrices de Ordenación Terriorial (DOT) de la Comunidad Autonoma del País Vasco. Euskadi, un Territorio hacia el Futuro[Z]. Victoria-Gasteis, 1997.

BILBAO

Marisa Carmona, Esteban Rodriguez Soto

The uniqueness of the site on a twisting estuary (La Ría) surrounded by mountains has become the scenario for city regeneration experience. Bilbao has embarked on a radical change of image after having been a mining enclave and the birthplace of the Spanish industrial revolution. In few years it has passed through an economic recession and is now a booming city. New, clean and modern industries are being created; the port is being rebuilt; major infrastructure and new transport systems have been done.

Many attribute this success to the visionary leadership of the city's various institutions, which have anticipated and developed several urban ideas and plans. They have taken advantage of opportunities offered by the crisis and have reoriented regional development towards its insertion in the European Union.

The city has generated a number of large urban projects through complex financial endeavours that have generated chain synergies, reversed the outflow of local entrepreneurs, created and improved accessibility and attracted national and international investment.

The Guggenheim Museum was one of the first strategic projects to be built along the Ría. The titanium building has become the driving force of the city's transformation.

Bilbao was characterised by its highly dense residential buildings which are only comparable to Hong-Kong. Between 1977 and 1988, it underwent a deep crisis and industrial restructuring started. Currently the unemployment rate has reduced slightly, and is probably the expression of the strong process of industrial reconversion and of the difficulties of the service sector to incorporate the labour surplus by itself. The idea was to promote culture; generate social and economic changes; enhance tourism, and to establish clean industries and services. All these have opened up new opportunities to the disadvantaged population dependent on the development of informal activities and small domestic and survival enterprises.

Architectural design has become a new element for city regeneration and employment creation. The titanium palace of Gehri, Calatrava's pedestrian bridge, Foster's metro access installations, the congress building Euskalduna of Soriano and Palacios, are all landmarks that have attracted a range of activities.

City History

The Villa of Bilbao was created in 1300 and several conflicts between the kingdom and the nobility influenced territorial arrangements during the Middle Ages. At the beginning of the 19th century, as in other European regions, Bilbao was characterised by administrative centralisation, the formation of the press, the development of trade unions, the formation of a Civil Engineers corps and the development of large industrial installations. Bilbao was also influenced by the Industrial Revolution, the ideas of the French Revolution, the nascent bourgeoisie, nationalism and ideologies of capitalism.

Two strategic projects were developed in the 19th century: the La Paz Port and the Ensanche both by Silvestre Perez. The largest urban change in Bilbao in the 20th century was made possible by an interconnected industrial operation in 1968. In fact, the great change was made a possibility when the petroleum refinery of Somorrostro decided to locate on the eastern coast of the estuary. In the tender for this civil works, the winning project introduced the idea of building the dike of Punta Lucero, giving the possibility to mooring large tankers. Thirty years later this decision brought considerable opportunities for the redevelopment of Bilbao, and the reuse of the industrial derelict area. New interventions such as the Abandoibarra, with the Guggenheim and Euskalduna, were also made possible.

During the first half of the 20th century several large industries were created. This industrial development produced significant wealth and full employment for many decades until the mid-Seventies. However the dependence of the productive structure on large traditional firms of mining, steel and shipping production produced the serious economic crisis of the Eighties. For more than 15 years the Basque economy deteriorated, and difficulties were encountered in integrating the region with the European Community.

In 1943 the first Metropolitan Plan for Bilbao included 21 Municipalities. The city grew with high density social housing developments, supported by large industries. A lack of land use regulations at the municipal level was noticed to guarantee spatial coherence. In 1956 the National Land Use and Ordering plan and in 1964 the Metropolitan Plan of Bilbao remained at a large scale (1 : 10000). Land use zoning was unable to adapt to the particularities of the territory and processes, lacking definitions at the local scale. The cities continue to grow surpassing urban regulations and the urban fabric became chaotic. This situation started to change with the enactment of new Land Use and Urban Ordering Law and Regulations (1978) and the transference of urban level competencies to the Administration of the Basque Country. With a new democratic constitution, Spain embarked on a period of empowerment of Municipalities that clarified public investments, and jurisdictional competencies for the different level and sectors. The General Plan of Urban Ordering of Bilbao (1998), and its regulations (DOT) are the new instruments that have guided the present changes.

The Impact of Globalisation

The Bilbao conurbation has become an important economic pole of technological articulation between the Atlantic Arc and the most dynamic spaces of Europe. This has helped to diffuse activities to neighbouring cities and to improve links to the corridors that connect the Basque Country. It is located in the corridor that runs from North Europe to Andalucia, through Paris and Madrid. Bilbao is currently focusing on technological innovation, infrastructure, and new management and marketing procedures. In the Zamudios Technological Park there are both high-tech large industries and state-supported intermediate R&D service-industries and small scale industries.

Two societies were formed to take advantage of this position and to react quickly to the opportunities opened up by macroeconomic realities:

● **The Society Bilbao Ria 2000** is made up of the Basque Government, the municipalities of the Ria, the Ministry of Development (Spain), RENFE (railways) Ministry of Industries (Spain), the Autonomous Port of Bilbao and the Foral Counties (Regional Government). The projects being developed by the Society are: the Railway Avenue, the intervention in the Amezola, the intervention in the Abandoibarra (Esukalduna and Museum Guggenheim), Interventions in Barakaldo.

● **The Bilbao Metropoli 30** is a public/private society aimed at creating synergies, to plan and foster future projects for the revitalisation of Bilbao.

Urban Strategic Planning

In Bilbao's planners are convinced of the importance of territorial and sectoral planning for achieving social goals. Studies have integrated territorial, morphological, visual, environmental, demographic and statistical components and resulted in comprehensive and social economic programming of the territory. Architectural expression and urban design (urbanism of ideas) are part of this analytical process and decision making. The urban projects should be presented to the people as demonstrative visual images of the future city. The main structural planning problems recognised are: the discontinuous process of development along the Ria, which gives a "lineal" expression to the agglomeration; the separation of both sides of the city and the need to build more bridges; the importance of the bordering elements; the underground system; the defence structures and the drawbridge structures amongst others.

The challenge of the new Guidelines of Territorial Ordering (DOT) is to enhance a territorial model that consolidates a system of cities to attract the economic, social and cultural innovations generated in the most dynamic spaces of Europe. The main belief is that the three important cities Bilbao, San Sebastian and Vitoria will be able to drive the whole Euskadi revitalisation in a balanced way. A coherent highway system has been planned to resolve existing inequalities and to enhancing economic development of secondary cities.

Strategic Plan

The Strategic Plan for the Revitalisation of Metropolitan Bilbao (1989) is the result of the organized consultation of various city stakeholders to facilitate a diagnosis, to establish a city vision and to specify the strategies required to make it a reality. The Metropolitan Plan aims to complete the basic infrastructure (Airport, Port, Railway system, Road System) to improve economic development and to maximise Bilbao towards national-level services and activities.

The City Territorial Plan has the following objectives:

● To integrate both banks of the Ría, and the generation of high quality urban spaces.

● To tackle the high density and overcrowding problems, especially on the left bank.

● To renew downgraded residential and industrial areas, through the creation of new economic activities and facilitating mixed residential, public spaces and services zones.

● To optimise the use of areas around the airport to create new developments, and an axis University-airport-Technological Park.

● To guide development in the coastal zone and create a system of municipal parks, including the recuperated riverbanks.

Urban Projects

Transport System

The connection with the Atlantic Arc is conceived as a basic European axis in which Euskadi has a predominant role. Internal communications must generate an efficient metropolitan network. The main criteria are:

Consolidation of a Basque Polynuclear System with a high capacity road system and a high- speed railway system with connection with Europe.

The development of high capacity public transport particularly the Metro of Bilbao.

The modernisation of the Port of Bilbao and substantial improvement to telecommunications and energy infrastructure.

Main Strategic Projects in Metropolitan Bilbao:

Planning and Integration of Infrastructure

A highway network and sanitation system; the underground has been upgraded along an old railway line and a new metro system on the left bank; the construction of the Abra Port, the Sondika airport, the technological Park of Zamudio, and extension of the railroad system towards the South, have contributed to the development of the Ria.

The Bridges

Bridges are an important element of Bilbao's history and identity. In 1893 a disciple of Eiffel designed and built the hanging bridge that joins Portugalete with Las Arenas. In 1997 as part of the Euskalduna, Manterola designed a covered radial bridge and Calatrava designed the Zubi Zuri, a pedestrian bridge. Frank Gehry integrated the Salve Bridge to the Guggenheim Museum and Fernandez the Deusto University Bridge.

The Metro

The Bilbao Metro, an old engineering project, was inaugurated in 1997, and has improved accessibility, urban productivity and the urban expression of the city through the creation of attractive entrances and stations designed by Norman Foster.

The New Port

A new Port with suitable installations in the Exterior Abra, has freed many areas that now are located in the urban centre for redevelopment. These new facilities are well connected with transport systems and with the supply of value added services allowing the port to be a major element in regional economic development.

The Airport

The Sondika airport (designed by Architect Calatrava), provides aerial connections in the Atlantic Arc, with extensive national and international services. A new modern airport building has been build, which provided service to 4 million passengers in 1999 and 10 millions in 2025.

The Barakaldo Exposition Building (Feria)

It is an emblematic project of national projection. Designed and built by Esteban Rodriguez Soto (SENER) and Cesar Azcarate Gomez (IDOM). There are 400000m^2 for exposition, parking and offices. The building has contributed to revitalize the Barakaldo Municipality.

Large Urban Project – Abandoibarra

The largest transformations of the Ria are taking place in Abandoibarra, which is an area of 346000m^2 in the heart of Bilbao. Until recently it was occupied by a shipyard, a large container station, two railway stations and a wharf. With the closure of the installations the sector underwent rapid downgrading. The revitalisation of the whole area has been managed by the Bilbao Ria 2000 Society, a public enterprise partly owned by the local authority (15%). Architect Cesar Pelli did the general urban plan. The organisation was constituted in 1992, with the city mayor as the president. The total program cost 180 millions dollars of which 26% was funded by the Bilbao Ria 2000 (through the sale of land to private concerns), 45% by the region and province (for the building of the Guggenheim Museum) and 29% by the Province and region (for the Concert Hall). Both the Guggenheim Museum and the Concert Hall Euskalduna are now functioning. The Guggenheim Museum (by Architect Frank Gehri) is a 25000m^2 building. The Concert Hall (by Architects, Dolores Palacios and Federico Soriano) has 25000m^2 with a main room for 2200 spectators, three small rooms, eight exercise rooms, and seven rooms for conferences and press meetings, cafeteria, restaurants and shops. The Centre for Leisure and Consumption Ria 21 (by Architect Peter Coleman) has 25000m^2 for shops and recreation. A luxurious Hotel of 14000m^2, the Pasarela of the University (Architect Jose Antonio Fernandez) a stainless bridge joining Abandoibarra with Deusto University, the Ribera Park of Architect Javier Lopez Chollet is a three km. park containing sculptures, will also contain the new building for the Rectory office of the Basque University, 800 new dwellings, and 81000m^2 of high standard office building, the new Library of the University of Deusto, a Maritime Museum and a grassfield Tram (with capacity for 250 persons).

A new development called "Olabeaga" is presently being developed by the team of Zadja Hadid. It consists of a new residential neighbourhood on the right bank of the river in the port-area.

海牙（荷兰）

图1　海牙在荷兰的区位图

案 例

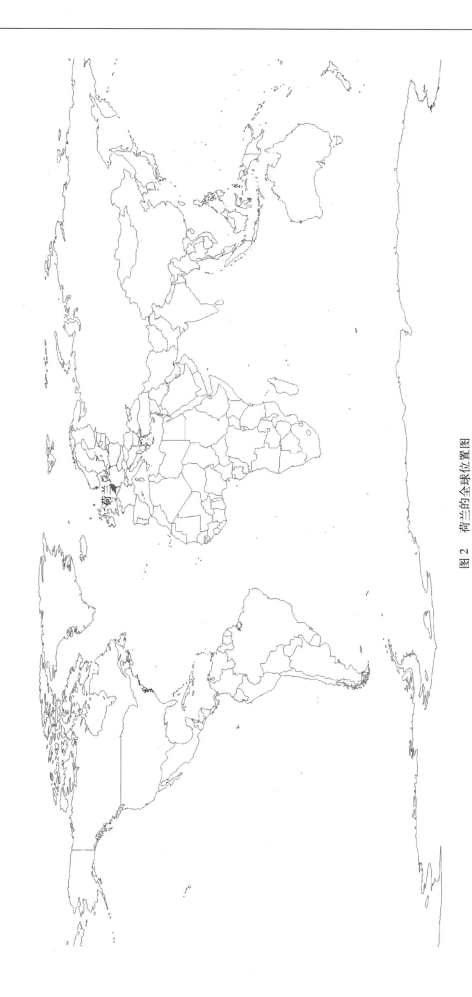

图 2　荷兰的全球位置图

海 牙

保罗·斯图登

一、城市数据

（一）海牙地区

没有都市圈。

人口：居民数量924785人。

城区数量：16个区。

经济增长率：3.6%，与荷兰全国平均水平持平。

（二）海牙市区

占地面积：8274hm^2（土地和水域）。城市化地区4281hm^2，55%为居住用地，18%为办公用地，27%为基础设施和服务用地。

人口：居民数量440729（1999），预计到2005年达到413473，到2010年达到408519。

1960年海牙有606000个居民。

1. 人口增长

与1960年比较，海牙人口增长大约为−27%，自从1985年起总人口稳定在440000左右。

2. 密度

总密度是103/hm^2（2001年），是除阿姆斯特丹之外人口密度最高的城市。

3. 就业人口

1998年工作人口为194151名。

4. 失业率

47%的人口没有工资收入（1997年），其中包括30%的老年退休人口，4.5%为接受无工作能力福利的人，12%为失业者或社会保障津贴受益人。市区以外其他地方35%的人口没有工资收入。

1998年有33186人没有或正在寻找工作。

住房拥有量：从1965年的174000套增加到1995年的204315套。2001年为213868套。

从1991年到2000年共新建18128套住宅。

住房类型：

家庭住房：14.7%。

有楼梯的多户住房：37.8%。

多户公寓：17.8%。

独立出入口的多户住房：9.5%。

公寓：7.7%。

其他住房：6.9%。

未知种类：5.6%。

（来源：BGB+）

住房户型分布：

1～2个卧室：13.7%。

3个卧室：22.7%。

4个卧室：35.7%。

5个卧室：12.8%。

6个或以上卧室：12.4%。

未知：2.6%。

所有权结构：

1994年私有住房比例为30%，市区外其他地方为44%，居住在租赁的社会住房占41%。

所有权结构（2001年，来源：OZB）

私有住房：37.8%。

公司拥有并出租：36.8%。

私人拥有并出租：21.5%。

未知：3.9%。

家庭大小：

户均人口从1965年的3.4减少到1995年的2.2。

住房短缺：

1992年有45000个家庭寻找住房，其中25%需求很迫切。

（三）教育，卫生

1. 教育

1996年海牙的学校数量：

VBO（小学-高中一贯中学）25所。

MAVO.C（高中）43所。

HAVO.C（技术高等院校）39所。

VWO.C（大学）42所。

2. 卫生

1998年有234名执业医生。

每1000个居民拥有汽车数量（1999年）：

—	私人汽车	两轮机动车	公司汽车	共计
海牙市区	331	12	53	397
海牙地区	366	17	51	435

二、城市概况

荷兰的中央政府、议会以及大量的国际机构与非政府机构，包括荷兰王室成员的住所都在海牙。另外，城中还有联合国国际法庭和战争法庭，是一个有着先进的服务、休闲、文化和旅游业的地方。由于这些品质，海牙在整个国家规划中具有特殊地位，特别是在南部的兰斯塔德地区的发展中。

由于荷兰土壤的特性，城市空间和社会人口的分布与不同土壤类型具有相关性（沙土或泥土地基）。一方面，王室府邸以及政府楼宇修建在沙土上，一些漂亮而绿化率极高的郡坐落在海滨，另一方面廉价公寓则建在含泥地土壤上，这些地方住着工薪阶层、小店主、生意人以及为那些郡里过着富裕生活的人服务而生存的人群，这些区域也是当前需要更新的都市区。第二次世界大战以后，海牙像其他荷兰城市一样，优先实施重建和更新。社会住房建设计划依据土壤的特征而进行，而不是社会和经济背景。现在不同区域之间的差异仍是由这些土壤结构的情况和补贴政策所造成的。

种族隔离情况比表面化的社会经济数据更严重，例如低收入或失业者，这看起来似乎是收入分化扩大的过程而不是城市各部分融合的过程。20世纪80年代国家的强劲干预和多项计划的实施已经产生显著作用，相对于美国的问题而言，两极分化的趋势已大大缓和。今天城市呈现出一定程度的社会异质性，并存在着相当的社会区隔。过去20年里，荷兰大城市的都市更新区已经得到了完全改造，在大量政府资金的支持下，海牙在城市面貌改善方面还是比较成功的，虽然社会经济上的差距仍然延续着。

从20世纪60年代以来，服务业在国民生产总值和就业中占据的比例逐年增加，并影响到了城市的全球化进程。从1965～1995年之间，服务人员就业在总就业人口中的份额从47%上升到69%，这是全球化的结果，也因为海牙更多地是一个行政性城市而不是工业基地，交通堵塞的情况也急剧上升（图3～图5）。

三、城市历史

海牙的城市结构体现了它的空间和社会结构，荷兰的国家管理机构及主要部门全部位于海牙。从一个13世纪的城堡发展而来，海牙逐渐发展成为一个重要的政治中心。15世纪由于政府机关和纺织产业增长而人口急剧增加。17世纪为了保护城市而挖掘了运河并确定了城市发展的模式。人口缓慢增长，至19世纪初，同时形成运河环绕下的开放结构。

注：图中显示了鹿特丹—海牙城市走廊（Stedenbaan），它和2008年建成的高速铁路站点一起加强了该地区的中心地位。

图3　包括雷登、海牙、代尔夫特、鹿特丹和豪达在内的兰斯塔德南翼地区

（来源：南荷兰省政府）

图4　航空照片
（来源：海牙市政府）

图5　火车站周围环境的模型
（来源：海牙市政府）

1850年后，城市在旧的市中心和荷兰Spoor车站（1843年修建）实现了首次扩张，并在Schilderswijk、Spoorbuurt和Rivierenbuurt规划了工人居住区，中产阶级则居住在Scheveningen往海滨的方向，1870～1930年之间人口从9.1万增加到43.8万。

当时的居住条件，特别是工人阶级区域的居住条件非常恶劣，对公共卫生及治安的担忧促使政府努力改善这些问题。在20世纪初期政府通过了住房法案（1901年）和公共卫生法案（1901年），从法律上帮助了私有机构从公共基金中得到经济援助来建造新的住房。

20世纪20年代，地方政府和住房协会建设了规模不大的社会住房，但直到第二次世界大战结束社会住房的建设都呈现减少的趋势，始于1913年向Scheveningen（坐落在海边区域）方向的城市发展仍在继续，为公务员和在印度尼西亚工作的人群而规划的住宅区绿化率很高。

经济发展与就业严重依赖于政府机构的存在，但海牙市仍发展为金融和贸易中心。由于20世纪前半部分的快速发展造成的混乱，城市进行了一些零星但缺乏整体规划的调整。相比鹿特丹和阿姆斯特丹等其他城市来说，海牙的自由主义力量更加强势，他们很不情愿接受严格的限制和总体规划，通过在个人私有

土地上限制自由修建豪华住所和乡村住房的手段来控制城市的发展。因为政治分歧,试图改变这种自由主义的态度和制定总体规划的尝试都失败了。然而1914年海牙却是第一个设立专门机构进行城市规划的城市,主要目标是控制工人阶级新区的环境状况和节制皇家及城市政府过度奢华的建设行为。

在第二次世界大战后的十年里,国家干预主要是增加住房供应量,并通过社会住房租赁来满足需要。在实施了多年交通规划以加强城郊与市中心的联系后,19世纪70年代的城市更新成为棘手的政治问题。与其他城市一样,城市更新关注的焦点主要是住房问题,海牙政府计划在城市更新区拆毁数以万计的房子。租客群体对房源减少与租金增加的抗争阻止了整个区域被破坏。因此与其他荷兰大城市相较,海牙采纳城市更新计划时有些犹豫。20世纪80年代时每年新建或改建的建筑平均超过1000座,20世纪90年代初上升到年均近1400座,1975~1995年间更新区域非常多。

1968~1995年间住宅增加了3万幢,并且总量达到了20.4万幢,而且住房产权的形式也发生了很大变化,社会租赁房的比例增加是通过政府大量收购私有住房实现的。这些房子或被改建,或被拆毁重建为新的社会住房。1971~1994年间,通过新建和私有租赁房出售的途径,自有住房的拥有比例成倍上升。

城市更新方法是以联合地方政府、租客组织和住房协会为基础的。这个战略叫作"邻里大厦",并且在住房和规划历史上是独一无二的。房客有优先权搬到新的或现代化的房子。分配和租金先事先制定好了,使租客组织积极参与规划和决策过程。直到20世纪80年代末,这些区域的房子只为社会部门而建立。

在20世纪80年代末,新住房和城市政策方法由议会批准了,被采取主要措施是:

(1) 从社会化住房向私有化住房转变以及刺激私人投资的规划政策导向;

(2) 从普遍性建设补贴向租金市场化后补贴最贫困的人群转变;

(3) 从公开地出租住房向住房市场化及鼓励个人拥有房屋转变。

十年多各区域各自为政的项目建设形成了拼凑式的格局,我们可以清楚地感觉需要更加整合的城市设计。除了需要与新的战略有关的设计工具外,过去几年有关市场方向和住房私有化的住房、规划政策改变已经引发了新的问题(图6)。

20世纪90年代荷兰的国家政策跟随分权、缩小国家对经济干预、私有化的全球趋势。城市更新采取根据战略规划的原则,并且更多注意集中在就业情况而较少地关注砖、灰浆和住房的特殊性。这些重建过程更多依靠市场力量,虽然当地政府必须对基础设施大量地投资并且承担保证这些计划连续性的风险。为城市扩张的特别计划会在所谓VINEX地方实行,在1998年,VINEX这个地方区域里面和附近的市区以及从毗邻自治市转移来的土地都归海牙的当地政府。同时规划合作和谈判已经增加了毗邻的15个自治市而一些为生存和未来经济活动的战略计划被发展来增强、复兴市中心(图7、图8)。

四、全球化,机遇和挑战

海牙是以四个大城市阿姆斯特丹、鹿特丹、海牙和乌得勒支为主的兰斯塔德大都市圈一部分。兰斯塔德缺少一个独立的权力机构,虽然它正在形成拥有自

图6 中部地区城市更新:社会住房
(来源:Paul Stouten)

己权利的大城市。在四个大城市以及三大都市圈（北翼-阿姆斯特丹地区；莱顿、乌得勒支；南翼-鹿特丹地区包括海牙）之间存在着竞争和合作。在对比伦敦和巴黎而言拥有过剩的办公空间和相对低廉的租金的时期，不同区域的办公住宅建筑正在热火朝天地建设，而且只有几个地区有机会发展成为有国际竞争水平的地区。在兰斯塔德大都市圈内，大量的国外及荷兰银行在阿姆斯特丹设立，而诸如石油服务公司之类的工程公司则在海牙和鹿特丹建立。更多的国际公司坐落在阿姆斯特丹和鹿特丹，其销售额的一大半是出口，而在海牙的企业则主要是与国内经济相关。

在20世纪70年代之前，大规模城市扩张和重建出现，高收入和中等收入群体开始郊区化的进程。同时来自南欧和北非国外的劳工出现汇集而且主要集中在中央商务区（CBD）附近的地方和城市边缘周围。20世纪70年代的外国移民继续归结于家庭团聚以及联邦公民流入量增加和苏里南与1980年代早期荷兰属地安的列斯群岛的独立。在1970年和1995年之间就业率下降了15%，就业损失特别发生在传统工业/公共服务业（-66%），建筑业部门（-65%）和贸易/旅馆/餐馆/修理部门（-50%）。但在服务部门有25%的产品服务和24%的其他服务的猛增量。第三产业在1970～1995年期间从47%增加到69%。然而办公空间建设增长只有全国平均水平的一半，这是由于可利用的地域短缺。自从1992年以来，这种情况伴随在市中心和公共交通系统节点附近的办公建筑建设刺激而发生了改变。

今天，海牙是在阿姆斯特丹之后的国家第二大行政城市。伴随服务产业的发展，对高素质劳动力的需求剧增。2002年海牙大多数高技能工作增量由通勤者提供，这从跨区域高速公路的交通堵塞就可以观察到。

在20世纪90年代初期，在199000个潜在的劳工中大约有27500人是失业的，2000年估计在50000～70000人之间，大城市失业率比全国平均水平还要高很多。这个情况是在就业结构中强烈的后工业

图7　根据城市扩张的 VINEX 政策计划：Ypenburg
（来源：Paul Stouten）

图8　社会住房与周边地区
（来源：Chiang Che-Sheng & Markus Vogl, 2005）

化转移的结果，集团存在是因为在劳动力市场缺乏一个牢固的位置以及缺乏来自顾客和区域激烈的竞争。

在1985～1995年之间，工资水平和社会福利之间缺口显著增加。在最低工资收入者的四个主要城市表现尤为突出。在海牙区之间存在着鲜明收入区别。因此可能几乎不会去争执在社会经济结构之内有鲜明对比。修建在沙土上的区域只拥有一个非常低的份额，大约只有10%～20%住房低收入群体。修建在泥煤土地上的工人阶级区域的份额是在50%～70%之间的。按收入和职业资格给予的两极化解释并不足以说明不同和不平等。津贴分配的政府措施使之更难在收入和对住房可及性以及公共服务之间建立一个直接关系。1991年政府大约重新分配65%国民收入——其中接近60%由家庭支付（税费、社会负担费用等）——

图9 在中心位置强化土地利用。附近建设的"议会大厦"
（来源：Chiang Che-Sheng & Markus Vogl，2005）

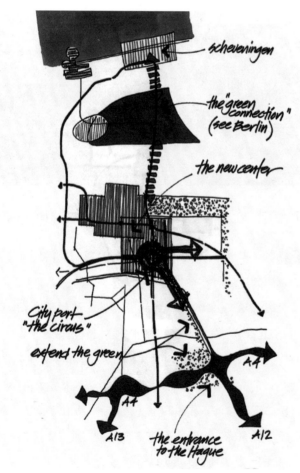

图10 战略计划的海牙与现有结构的关系
（来源：南荷兰省）

主要在住房、教育、文化和文娱基础设施方面。但高收入家庭级别仍然比低收入家庭级别支付得更多。房主特别是那些高收入者能从中得益，因为他们能从贷款中得到利息补偿而且没有最大额度限制，这是在欧洲独一无二的。在海牙，房东的收入水平和收入预期都远远高于租房的人群。此外，族群分离明显，这些族群的最大聚集地通常在人口密度很高但没有工作的传统战前城市更新区域中被发现——超过20%的比例——在两个城市更新区，而且失业率在20世纪50年代修建的战后区大约为城市总量的平均水平。

在20世纪80年代更多关注投向城市更新区，而郊区则慢慢减少。在20世纪80年代末，经济复苏和住房计划开始从社会向市场部分转移了。更多的关注投向环境和流动性的消极作用上。办公楼和豪华公寓的新计划已经开始发展，这是为了市中心的关键地域和城市扩张（图9）。有关当高和中等收入的团体搬离这些旧区域而把人口聚集的低收入团体扔在一边是否会增加族群数量的争论不断。同时也发生了来自亚洲、非洲和东欧的避难所寻找者的涌入。

1995年中央政府和海牙政府签署了一项关于通过大城市政策来防止生活质量和公共安全降低的协议，同时有四个主题活动，包括了工作，安全，关爱和教育。一项重要措施就是介绍些特别计划来为那些至少一年没有工作以及靠社会保障津贴生活的人创造工作。这些在几个城市更新区计划将和物理计划相联合来击退失业以及在少数族裔团体和青年人之中促进雇员增加量。那些计划关注公共空间的安全、维护及改善来改进生活质量水平。最后目标是"走向一个统一的城市"（图10～图12）。

五、城市战略

从20世纪90年代城市政策发生了根本变动。城市再生政策是整体城市和城市区域合并计划的一部分，并且改进城市竞争性的经济目标开始作为主要事件。这种新方法与政治和经济发展的变化相连接，这是新背景下的私有化和解除干预的重要组成部分。建筑环境与住房的维护与管理不再被视为需要中央政府补贴的任务，而是建筑所有者（包括住房组织及地方政府）的责任。国家空间组织的变化在市政司法之外提供了城市地区的概念来处理城市扩张问题。大都市政策瞄准了为城市化以及城市及附近建设计划提供指南来避免次级都市化的消极客观性，主要是能源的利用。紧凑城市的方法变得很重要。加强社会经济容量被看做预言目标，改变对老城市区的掠夺情况以及改进住宅在国际上可接受标准的需求关系。

在20世纪90年代中期，中央政府做出重要一步来削减建设补贴和住房预算，这几乎接近消失在国家预算外。在1992年，关于为将来（Belstato）城市更新政策的备忘录里，总共有6180万英镑的投资被确认为到2010年城市更新的需要。在1996年1月1日海牙接受了中央政府4325万的补助来解决问题（土地污染和纪念碑被拆除）并且海牙自己本身捐献了共计4600万，但大多城市投资不得不由私人投资者来实现。

市政工程

1990年在荷兰城市一些项目被认为是关键项目由公私合作（PPPs）来实现。ABP是在这些计划中十分重要的投资者之一，它是公务员退休金基金。

那些项目选择标准包括：

（1）刺激国际联系来开发在兰斯塔德的主要用地，鹿特丹机场和阿姆斯特丹Schiphol机场，运输主轴、都市结以及把荷兰作为水乡的想法；

（2）对收缩公共交通津贴易变性有贡献的项目以及减少住宅工作地点之间距离的那些项目；

（3）带有空间措施和环境政策的那些项目；

（4）改进经济结构和区域或城市生活水平并且应付大规模问题的那些项目。

在海牙，战略计划覆盖着六个主要地方，它们包括：

（1）在中央驻地附近的区域"新城市中心"这个中心以办公发展为主，与第二部分为办公、购物、酒

图11　城市更新促使下的土地用途变化和土地集约加强
（来源：海牙市）

图12　创建一个新的中心地位区连接到中央站
（来源：Chiang Che-Sheng & Markus Vogl）

店和高收入群住所的Beatrix Quartier相连接。所有与新Randstadreil相连接的是在兰斯塔德南翼连接城市和村庄的跨区轻轨。

（2）由理查德·迈耶设计的新城镇厅大楼和地方政府办公室以及市内地下转换结点方式。

（3）荷兰Spoor区车站，先前为老工业区，将被改变为商务、办公发展和多工艺教育中心，包括连接南部地区的车站改善。为教育目的保留的80000m²土地。

（4）更新区的老市中心的商店，旅游和文化设施。

（5）Scheveningen港口和为旅游，钓鱼和乐趣寻

找的海边胜地。

(6) 荷兰会议大楼附近，给机构，会议，画廊和旅游的区域。

另一方面，为促进生活水平、安全和城市经济，海牙在 1996～1999 年之间接受了 1150 万欧元，欧共体为城市更新计划提供 1200 万来改进就业、教育、公共安全和在 Schilderswijk 区的多元文化事业。

未来有着许多不确定性。更多设计和规划手段，以及更多企业用地、交通计划和基础设施是必需的。随之而来的问题是，从这些投资和产业结构调整中受益的人是谁？看起来城市更新计划更有利于高收入阶层，而对于低收入者的补助却在减少。而物质性空间、社会、文化和环境要素之间的联系未在城市规划实践中得到关注。

接下来十年的主要项目是在中央驻地附近，这些地方成为连接城市经济增长和城市结构整合的关键地方。这个区域成为兰斯塔德南翼新区基础设施的中心节点。发展成为国际顶级地区的雄心将由包括办公室、商业中心、休闲设施和豪华公寓在内的混合计划所支持。同时也有一直持续到 2010 年的十分重要及相当数量的住宅计划。虽然混合住宅想法存在，但大约 15600 套低成本住宅计划由 11000 套中高档收入住宅所代替（图 13～图 19）。

图 13 中央区附近的居住区的航空照片
（来源：Paul Stouten）

图 15 海牙中心站和苏黎世塔
（来源：Chiang Che-Sheng & Markus Vog）

图 14 海牙中心站和苏黎世塔
（来源：Chiang Che-Sheng & Markus Vog）

图 16 中央站周围的土地集约利用
（来源：Chiang Che-Sheng & Markus Vog）

图 17　创建的新建筑在海牙市的再生
（来源：Chiang Che-Sheng & Markus Vog）

图 18　苏黎世塔
（来源：Chiang Che-Sheng & Markus Vog）

图 19　该市的地铁列车站
（来源：Chiang Che-Sheng & Markus Vog）

六、大规模城市开发项目

滨水区战略计划的居住部分

1. 背景

几个世纪以来这是海牙内地航海港口区。19 世纪末，在这个城市的中心地点一些运河仍然在建造。大约在 1900 年，这个区域的生活环境非常恶劣，部分是由工业化进程导致的变化所引起的。1908 年第一个重整结构计划由 Berlage 设计。这个计划没有实现，只完成了几个街区与角落。第二次世界大战的破坏和 20 世纪 30 年代经济危机导致延迟了这个区域的变革。在 20 世纪 60 年代一位私人开发商委托意大利建筑师 Nervi 为他设计了一座 140m 高的摩天大楼，但最高法院（Raad van State）拒绝了这个提案。

20 世纪 80 年代，新的城市设计进程开始了。第一想法来自建筑师 Carel Weeber，包括创造一个步行区域结束城市内部汽车交通堵塞的弊病。这个计划打开了未来发展之路和中央政府为住房计划部门和中央驻地附近的环境修建一个新大楼作出贡献。

这个区域的第二个做法就是决定在这个区由当地政府来修建新市政厅。这个地方的主要争论是，新市政厅的建设应该成为吸引私人投资者参与该区域建设现代化市中心的信号。新市政厅的设计选择了知名建筑师理查德·迈耶。市政厅是一个多功能大楼，结合了民事用途的图书馆、餐馆、饭店、商店和商业办公室，提高了人们的可及性。下一个计划是为农业部修建一个新大楼。

2. 居民区作为新中心一部分

在内阁成员拒绝建造农业部和渔业部大楼的计划后，海牙的银行区开发包括居民区新中心的新动议被采纳。三个主要土地拥有者——海牙的自治市、中央政府和荷兰铁路局作为这个新动议的代表（Spaans, 2000）。1989 年新中心城市工程达到了一个关键的阶段，其作用得到了海牙自治市和当地政府的住房规划环境部门、内务部门、经济事务部门以合同方式的确认。随后工程满足了全国空间调整的建筑容量与土地使用要求（IV 和 VINEX 部分），因此通过工作与居住地点的集聚而降低了私人汽车的使用需求（Nijpels 等，1988 年；Alders，1991）。

修建大城市中心的目标是获取国际范围的竞争力。为达到这个计划，也提出了一个混合功能的计划，包括企业设施、休闲功能、高标准住宅环境、豪华商店和娱乐作用（如由著名建筑师库哈斯设计的音乐舞剧院），一个新图书馆。计划包括对公共交通可及性的提升和高质量公共空间的改善。这个工程吸引了很多人的注意，包括：当地政府、退休基金会（ABP0）、全国铁路公司（NS）、中央政府大楼建造部门（Rijksgebouwen Dienst）、本地交通公司（HTM），以及几个大零售公司（像 V&D, Bijenkorf, HEMA 和 C&A）。根据这个计划，200000m^2 的办公楼，120000m^2 的商店、餐馆、酒吧和公共设施以及 1200 套住宅将会实现。通过修建一个满足公共交通需求的新隧道，地下空间也将得到改进，并且新兰斯塔德轻轨和中心车站将被扩大，同时也扩大为汽车交通和新停车处设施的新轨道和隧道。

这个项目的总投资（除了中心车站）估计为 12 亿欧元，其中 10 亿欧元将会来自私人投资者。快速电车隧道投资需求估计为 1.136 亿欧元（包括两个地铁站），汽车交通隧道为 6360 万欧元。简而言之，自 2000 年初始，超过 22.7 亿欧元投资用来更新城市中心部分。

3. 居住规划指导方针

1996 年城市发展部门改组确定，并且一个新的特别负责站点管理、项目发展和土地事宜的部门建立了。伴随这些变化，当地政府意欲更好装备开发战略计划。在改组以后，只有一位市议员代替原来两位来负责新中心和居民区。在整个进程中合作伙伴遭遇了些变化。原有的合作伙伴包括地方政府、中央政府、荷兰铁路公司以及公务员基金，后来又加入了私人开发商 MAB（担任项目协调人）。在海牙自治市收购土地以后，土地卖给了私人开发商。在提出第一个设计概念以前，荷兰铁路公司退出。之后退休基金会退出，因为居住建筑中出租房份额改变进而投资价值减少了。之后，私人开发商 MAB 决定与其他私人开发商一起加入合作关系。

4. 居住区设计

1987 年建筑师罗布·克里尔被委任进行城市规划设计。第一个提议是在市中心和中心车站之间建造一条步行路线，并作为设计原则的一个重要方面。为了工作的便利，还专门成立了一个工作组用来进行设计活动并在几个（可能的）伙伴之间沟通。这个工作组由政府建筑师代表的中央政府和议员所代表的地方政府所组成。除罗布·克里尔以外，另外还邀请了四位荷兰建筑师。在工作组里讨论的重要设计议题包括街区的建筑密度、地下电车轨道、精确的定位和中心广场的形式，与毗邻街区和现有建筑联系的入口。在工作组里，私营部门变得热心并且外国建筑师也被委任设计建筑项目。除六位荷兰建筑师以外，西萨·佩里（美国）、迈克尔·格雷夫斯（美国）和 Adolfo Natalini（意大利）与罗布·克里尔一样任建筑开发项目监督员，Soeters 委任为日常监督员。在这个项目中会实现以下内容：115000m^2 办公室，315 所住宅，800 个停车场，

2000m² 为自行车停车处，4500m² 包括服务的商店，54000m² 办公室为住房中央政府机关。

5. 居住区建筑项目

这个塔楼包含23100m²办公室和88m高的19层楼。建筑师是西萨·佩里。

Muzentower 是八角形形式，是73m高的17层楼。这是居住区主入口的一部分，在新城镇厅对面。塔楼有 13500m² 办公室。

其他办公楼中的一座有有轨电车跨过并与商业设施连接。这个大厦的建筑师是罗布·克里尔。

这个大厦以前是该市印刷办公室，现代化改造后它用作教育作用以及停车场入口。

七、结论

整合工作和靠近公共交通连接点的住宅区的目标已经实现了。因为在办公室层面有更多的力量和更多的收益，所以在整个过程期间相当数量的住宅减少了。全球化的影响不仅在于审美和形象因素重要性以及国际标准，而且因为著名建筑师的介入和很多国际私有资本的财政结构。大约50%居住区是由德国投资基金提供经费并且最后拥有。建筑计划在几个部门传播意味着传播和扩展风险以及给实现计划带来灵活性。为了控制品质而实行了新的机制，如由建筑师与项目监督员、政府官员和其他工种专家组成工作室。除公共私人合作关系以外，新管理和规章制度将借助重组城市发展部门而转向一种更具市场敏感度的方向发展。

参考书目

[1] Alders J. Vierde Nota over de Ruimtelijke Ordening Extra, deel III；Kabinetsstandpunt. Den Haag；SDU Uitgeverij. (Fourth Memorandum on Planning Extra)[Z], 1991.

[2] B & W Gemeente Den Haag. Investeren in Vernieuwing；Nota Stadsvernieuwing[Z]. Gemeente Den Haag (To Invest in Regeneration；Memorandum on Urban Renewal), 1994.

[3] Dienst Ruimtelijke en Economische Ontwikkeling. Structuurvisie Den Haag[Z]. Deel 1：Hoofdlijnen；Gemeente Den Haag (Urban Development Vision The Hague；Part 1：Main Issues), 1994.

[4] Dienst Ruimtelijke en Economische Ontwikkeling. Structurvisie Den Haag[Z]. Deel 2：Achtergronden en uitwerking；Gemeente Den Haag (Urban Development Vision The Hague：Part 2：Background and Elaboration), 1994.

[5] Dienst Stedelijke Ontwikkeling. Stadsanalyse：De Nieuwe Kaart van Den Haag[Z]. Gemeente Den Haag (Urban Analysis：The New Map of The Hague), 1998.

[6] Dienst Stedelijke Ontwikkeling Directie Wonen. Wonen in Den Haag；verscheidenheid, vitaliteit en duurzaamheid；Stedelijk beleidsplan Wonen；1998-2001[Z]. Gemeente Den Haag (Living in The Hague；Diversity, Vitality and Sustainability), 1998.

[7] Gemeente Den Haag. Een Herstructuringsplan voor de 20e eeuw；Naar een Ongedeelde Stad[Z]. A Restructuringsplan for the 21th Century：towards an Undivided City, 1998.

[8] Kruythoff H., B. Baart, W. van Bogerijen and H. Priemus with J. den Draak. Towards Undivided Cities in Western Europe[M]. Part I：The Hague. Delft：Delft University Press, 1997.

[9] NS-Vastgoed：Den Haag CS-Masterplan；Amsterdam[Z].

[10] Nijpels E., R. Lubbers. Vierde Nota over de Ruimtelijke Ordening, deel d：Regeringsbeslissing[Z]. Den Haag：SDU Uitgeverij (Fourth Memorandum on Planning), 1988.

[11] MVROM (Ministry of Housing, Planning and Environment). Workshop La-Vi kavel[Z]. Den Haag, 1991.

[12] Spaans M. Realisatie van Stedelijke Revitaliseringsprojecten；een Internationale Vergelijking[M]. Delft：Delft University Press, 2000.

[13] Stouten P. Divisions in Dutch Cities[J]. City, 2000, 4 (3).

THE HAGUE

Paul Stouten

The Hague is part of the Randstad conurbation in which four large cities Amsterdam, Rotterdam, The Hague and Utrecht are dominant. There is competition but also collaboration between the four large cities and also within each of the three conurbations (North-Wing, the Amsterdam Region; the strip Leiden, Utrecht, and the South-Wing, the Rotterdam Region including The Hague). There is an increase in the construction of office accommodation on these different regions, there are periods when there is a surplus of office space and relatively low rents compared to London and Paris. Moreover only a few locations have the chance to develop into top locations, which can compete at an international level. Within the Randstad conurbation a significant number of foreign and Dutch Banks are established in Amsterdam whilst engineering firms, for example, serving the oil companies are to be found in The Hague and Rotterdam. Companies in Amsterdam and Rotterdam are more internationally oriented. Over half of their sales are exports whilst in The Hague (1 : 25) businesses predominantly operate within the national economy.

Before the 1970s large-scale urban extensions and reconstruction appeared and suburbanization of higher and middle-income groups started. At the same time the influx of foreign labour from southern Europe and North Africa took place, which was mostly located near the Central Business District (CBD) and in the urban fringes, the Periphery. Between 1970 and 1995 employment declined by 15%. The employment loss occurred particularly in the traditional industries/public services (−66%), the building sector (−65%) and in trade/hotel/restaurants/repair sector (−50%). But in the service sector there was a strong increase of 25% for production services and 24% for other services. In the period from 1970 to 1995, the tertiary sector increased from 47% to 69%. Nevertheless office space construction grew by only half of the national average, due to shortages of available locations. Since 1992 the situation has changed with the stimulation of office construction in the city centre and around nodes of public transport systems.

Today The Hague is the second most important office city in the country after Amsterdam. Together with the rise of the service sector there has been an increase of the demand for qualified labour force. Most of the increase in professional jobs in The Hague has been filled by commuters- approximately 44% of the total in 1995, which is to be seen in traffic jams of interregional highways. At the beginning of the 1990s about 27500 people out of a potential workforce of 199000 were unemployed. In the year 2000 it is estimated that total will be between 50000 and 70000. Unemployment rates in big cities are substantially higher than the national average. This situation is the result of the strong post-industrial shift in employment structures; the presence of groups lacking a strong position on the labour market and strong competition from commuters from the region.

Between 1985 and 1995 the gap between wage levels and social benefits increased significantly. In the four major cities, households which belonged to the lowest wage earners were over-represented. Sharp income differences existed between districts in The Hague. So it can hardly be disputed that there are sharp contrasts within the socio- economic structure. But an interpretation of social polarisation exclusively in terms of income and occupational qualifications gives an inadequate image of differences and inequalities. Government measures for the allocation of grants have also made it more difficult to establish a direct relationship between incomes and the accessibility to housing and public services. Segregation of ethnic groupings is strong.

In the eighties more attention was paid to the urban renewal districts while suburbanization slowly decreased. At the end of the eighties economic recovery started and housing programs shifted from the social to the market sector. More attention was given to the environment and to the negative effects of mobility. New plans for office building and luxury apartments have been developed for strategic locations in the city centre and for urban expansion. Concern has arisen about whether these plans increase segregation when higher and middle-income groups move out of the older districts and leave higher concentrations of lower income groups behind. At the same time an inflow of asylum-seekers from Asia, Africa and Eastern Europe has occurred.

In 1995 the central government and the local government of The Hague signed agreements on metropolitan policies to control the decline in the quality of life and public safety with five themes for action amongst which are: work, safety, care and education. One important measure is the introduction of special programs to create jobs for people who have been out of work for at least one year and living on welfare benefits. These programs in several urban renewal districts will be connected with physical planning to control unemployment and promote recruitment of employees among ethnic minority groups and young people. Those plans give attention to safety and also to the maintenance and improvement of public spaces in order to improve the quality of life. The ultimate target is 'towards an undivided city'.

Urban Strategy

Since 1990s urban policies have undergone fundamental changes. Urban regeneration policies are part of an integrated planning of the whole city and city region, and economic targets to improve city competitiveness started as main issues. This new approach is linked to changes in political and economical development, which included privatisation and deregulating important components of the new context. The maintenance and management of the building environment and the housing stock were no longer seen as tasks to be subsidised by the central government rather than responsibility falling on owners and local governments. Changes on national spatial ordering provide the concept of city regions to deal with urban extensions beyond municipal jurisdictions. Metropolitan policies were aimed at providing guidelines for urbanisation and construction programs in and around cities to avoid negative externalities of sub-urbanisation, mainly energy use. Compact city approaches became important. Strengthening of social-economic capacities was seen as a target to predict and reverse the deprivation of old city districts and to improve residential relations to required internationally acceptable standards. In the mid 90s important steps were made by the central government to cut back construction grants and housing budgets, which virtually nearly disappeared out of the national budget. But most of the urban investments have had to be realised by private investors.

Urban Projects

In 1990 several projects in Dutch cities were recognised as key projects to be realised under Public Private Partnerships (PPPs). One of the important investors in these programs is the ABP, a pension fund for public servants.

Selection criteria for projects included:

— Those which stimulated international connections with projects to develop metropolitan locations in the Randstad, the Port of Rotterdam and Schiphol Airport, the main axes of transport, urban nodes and the idea of the Netherlands as a water country;

— Those which contributed to shrinking mobility with grants for public transport and which decreased the distance between residential and work places;

— Those with spatial measures and environmental policies;

— Those which improved economic structures and quality of life of the region or the city and which dealt with issues at a larger scale.

The future holds many uncertainties. The problem, of which population groups are benefiting from these investments and changing employment structure, remains. It seems that urban regeneration programs are oriented more to the upper income side of the housing market and grants for urban renewal are decreasing. The connection of physical-spatial, social, cultural and environmental aspects is generally not pursued in urban planning practice, and is in fact, never implemented. The main projects for the following decades are around the Central Station, which has become the key location for linking the growth of the 'urban economy' and city 'restructuring'. This district became a central node in the new regional infrastructure of the south wing of the Randstad. The ambition to develop an international top location will be supported by mixed programs including offices, businesses centre, recreation facilities and luxury apartments. There is also an important amount of dwellings plans till 2010. Ideas on mixing dwellings exist, however about 15600 low cost dwellings are planned to be replaced by 11000 dwellings for middle and higher income groups.

Large Urban Project

The Resident part of the strategic plan for the BANK district

Background

In 1980s new urban design processes started. The first ideas came from the architect Carel Weeber, consisted of creating a pedestrian zone to put an end to the dominancy of the car traffic in the inner city. This plan opened the way for further development and the central government contributed by building a new building for the Ministry of Housing Planning and Environment near the Central Station. The second intervention in this area was the decision of the local government to build the new town hall in this district. One of the main arguments of this location was that, the new town hall should become a detonator for the development of other plans in partnership with private developers as a way of developing a modern city centre. For the design of the new town hall, a well know architect Richard Meyer was chosen. The town hall is a multipurpose building combining civic uses as well as library, restaurants, pubs, shops and commercial offices to enhance the accessibility for the population. The next idea was to build a new building for the Ministry of Agriculture.

The Resident as part of the New Centre

After plans for building the Ministry of Agriculture and Fishery by cabinet were rejected, new initiatives were taken to develop the new centre of The Hague for the BANK district including Residential areas. The three main landowners, the municipality of The Hague, the central government and the Dutch railway company were represented in this initiative (Spaans, 2000). In 1989 the urban project for the new centre reached the status of an officially recognised key project which was confirmed in a contract between the Ministry of Housing, Planning and Environment, Ministry of Home Affairs, Ministry of Economical Affairs and the local government. The project met the requirements regarding the building rate capacity and land uses as established by the National Spatial Ordering (IV Note and VINEX) to discourage private car use through enhancing work and living concentrations near transport nodes (Nijpels et al., 1988 and Alders, 1991)

The goal was to construct a metropolitan centre able to compete on an international scale. For doing this a program with mixed functions including business facilities, recreational functions, a high standard residential environment, luxury shops and recreational functions, a new library was also proposed. The program included the improvement of accessibility to public transport and high quality public spaces. In this project much interest was shown by the: Local government; Retirement funds (ABP), the national railway company NS, the central government department for governmental buildings, the local transport company HTM, and several big retail companies like V&D, Bijenkorf, HEMA and C&A. According to the program 200,000m^2 of office building, 120,000m^2 of shops, restaurants, pubs and public facilities and 1,200 dwellings will be realised. The infrastructure will also be improved by constructing a new tunnel for public transport and the new 'Randstad Light Rail' and the central station will be enlarged with new track, tunnels for car traffic and new parking facilities.

The total investment of this project (excluding the central station) was estimated at €1.2 billion of which €1 billion would come from private investors. The investment required for the tunnel for the speed tram, was estimated at €113.6 million (including two underground stations), and a tunnel for car traffic at €63.6 million. In short, at the beginning of 2000, an investment of €2.27 billion to renew the heart of the city was a fact.

The Resident Planning Guidelines

In 1996 a reorganisation of urban development department was settled. After the reorganization only one Alderman instead of two was made responsible for the New Centre and Resident district. The partnership suffered several changes in the whole process. The original partnership between the corporation of local government, the central government, Dutch Railway Company and the pension fund for civil servants was extended to the private developer MAB, which was commissioned as project coordinator. After acquisition of the land by the municipality, the land was sold to the private developer. Before the first concepts of design were drawn, the Dutch railway company retired. Later the retirement fund was withdrawn because the investment value was reduced when the share of rental housing changed in owner occupied housing. Later the private developer MAB decided to enter into partnership with another private developer.

The design of the Resident

In 1987 Architect Rob Krier was commissioned to design an urban plan. The first proposal for creating a pedestrian route between the city centre and the Central Station was an important aspect of the design guidelines. A workshop was used as an instrument to facilitate the design activities and negotiations between several (possible) partners. During the workshop important design themes were the building ratio of the blocks, the underground tram track, the precise orientation and form of the central square, the entrances of The Resident from the adjacent districts and the adaptation of existing buildings. With the workshop, the private sector became enthusiastic and also foreign architects were commissioned to design the building projects. Besides six Dutch architects, Cesar Pelli (USA), Michael Graves (USA) and Adolfo Natalini (Italy) were commissioned to develop the building projects with Rob Krier as supervisor. The following programme was realised in this project: 115000m^2 of offices, 315 dwellings, 800 parking places, 2000m^2 for parking bicycles and 4,500m^2 of shops including services. 54000m^2 of offices was for housing central governmental institutions.

Conclusions

The main goal of integrating work and residential areas near public transport junctions has been achieved. During the process the amount of dwellings decreased since the claims on office surface had more power and are more profitability. About 50% of The Residential areas were finally financed and owned by German investment funds. The construction plan spread in several sections meant spreading out and broadening the risks and it gave flexibility in realising the programme. For developing quality control, new instruments were used like a workshop with architects and from the commissioner of works to a general supervisor, 'the authority'. Besides the Public Private Partnership new instruments of governance and regulations were developed by reorganising the department of urban development towards a more market sensitive approach.

基多（厄瓜多尔）

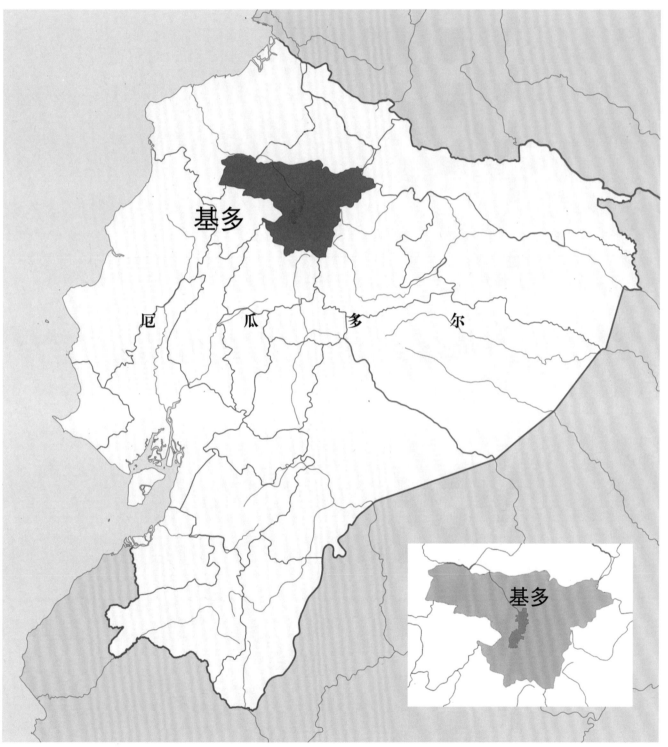

图1　基多在厄瓜多尔的区位图

案 例

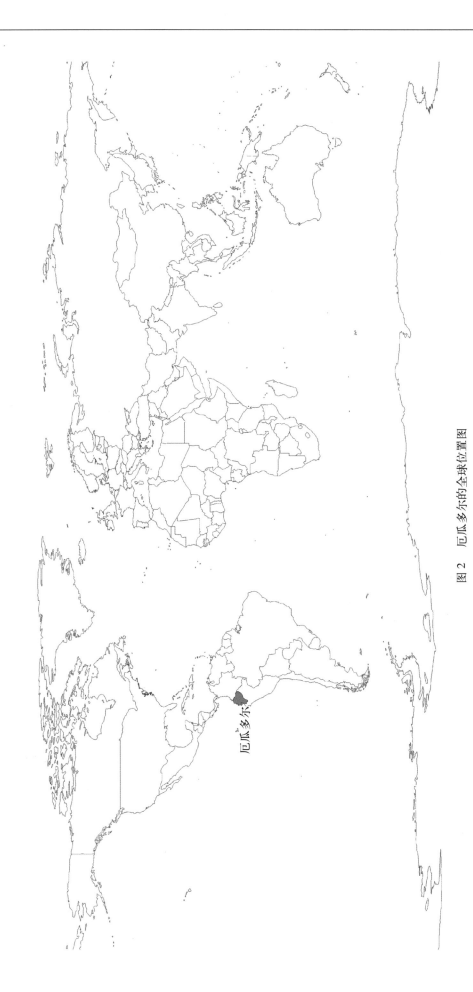

图 2 厄瓜多尔的全球位置图

基 多

迭戈·卡利翁

一、城市数据

（一）基多都市圈

面积：4204km²，其中生态保护区：1922km²。
农业用地：1944.42km²。
人口：1842201 人。
密度：438.2 人/km²。

（二）基多市

面积：324km²。
人口：1397698 人（2001）。
人口增长：1.6%。
密度：4313.9 人/km²。
活跃就业人口数：715415 人（1999）。
失业率（1999）：11.30%。
非正式就业率（1999）：31.10%。
就业分布（1999）：
社会服务业：36.39%。
零售及餐饮业：26.44%。
制造业：17.98%。
金融服务业：8.44%。
交通与通信：4.39%。
建筑业：4.37%。
农业：1.16%。
采掘业：0.77%。
不确定职业：0.05%。
非正式就业：24%。
城市财政状况：
支出（2001）：108371900 美元。
债务：5023261 美元。
投资：43700000 美元。
其他：8700659 美元。
收入（2001）：108371927 美元。
税收：28309374 美元。
服务：10664100 美元。
自有资源：13878500 美元。
转让收入：52518453 美元。
其他：3001500 美元。
年人均税收：51 美元。
住房状况：
住房数量：363997 套。
家庭数：86666 户。
住房种类：
独立住房：42.30%。
公寓：29.60%。
客房：20.60%。
其他：7.50%。
住房权属：
自有：37.90%。
租赁：46.40%。
其他：15.70%。
贫民窟数量：25 个。
非法住区数：200 个。

住房短缺数：20602套。

市政设施覆盖率：

自来水：93%。

排水：82%。

供电：96%。

电话：55%。

交通设施：

街道长度：1300km。

轨道交通：240km。

年公共交通乘客数：1.095亿人次。

年机场乘客数：120万人次。

文化设施：

博物馆：36个。

艺术廊：64个。

电影院：21个。

剧场：14个。

图书馆：99个。

教育：

幼儿园：594所。

小学：837所。

中学：452所。

大学：18所。

合计：1883所。

卫生保健：(1998)。

健康中心：86所。

诊所：37所。

医院：17所。

病床数：4586张。

药房：640间。

娱乐：

餐厅：1229家。

咖啡厅：65家。

市民体育：

游泳池：15家。

公园：1331个。

人均绿地面积：8.15m²。

足球场：92个。

年固体垃圾数量：474500吨。

二、城市概况

基多是厄瓜多尔的首都和国家中北部地区的中心。城市位于南北走向的安第斯峡谷，海拔2820m以上，城市西边是皮钦查火山，东边是伊钦比亚丘陵。由于平地稀缺、群山环绕造成地形不规则，以及无数的东西向斜坡，这些条件一直制约着这个城市的建设。

基多是一个重要的移民目的地，在过去几十年中经历了显著的人口增长，1980~1990年间每年人口增长达到3%。目前人口150万，占全国人口总量的12%，占全国经济活动人口的16%。基多还占有全国制造业总量的30%和公共服务总量50%的份额。（来源：Propuesta MDMQ-DGP 1996. Régimen Distrital del Suelo, Quito. p.15.）近20年，基多地区经历了重要的空间转变，城市区域从"中心导向"的城市开始发展，伴随小的城市中心和周边农业地区的合并，发展成为覆盖了Tumbaco-Cumbaya峡谷、Los Chillos峡谷、Calderon和Pomassqui-San Antonio de Pichincha峡谷的一大块区域（图3、图4）。

城市规划对其形象和城市的空间特征产生了重要的影响。其物质性规划过程开始于20世纪30年代末，Jones Qdriozola的规划方案包括了面积为3376hm²的

图3　卫星图

土地（1942年），基多大都市圈区域规划方案（1993年）和它的土地使用规划方案（1995年）范围19014hm^2，2020年的基多战略规划包括土地42472hm^2（1998年）。

规划过程表面上由技术理性推动，但实际上是服务于"既成事实"，以及法律规定的现实情况或者政治经济利益。在大多数情况下，不足之处在于土地使用导引和规章制度虽为此目的而制定，但执行起来难度却很大。自1993年起，基多大都市地区法提出了比城市法所规定的传统城市职能更加广泛的法律框架。以下是城市法为所有市政当局在市立法律的基础上增加的土地使用权力：

（1）界定生态保护和环境保护区域的强制性权力；

（2）土地使用转让方面的强制性权力；

（3）土地管理协调方面的强制性权力；

（4）交通规划管理上的强制性权力（由于交通条件对土地或房产带来的或促进或不利的影响）。

基多大都市地区法是一个仅仅适用于基多的城乡管理工具，在城市中心疏解和分散管理的过程中，这部法律对大城市区域管理方面的变化产生了重要影响。

三、城市历史

基多市最初是基多国的印第安部落，后来被印加人占领，随后又被西班牙人征服并作为殖民地。Sebastián de Benalcazar 在1534年建立了基多，从西班牙式政治到经济的殖民管理对这个城市及其功能都进行了重组。西班牙式城市是基于罗马的方格网格局而建立，有着强烈的社会居住隔离性：西班牙和 Criolla 人在城市中居住，印第安人则居住在城市外围。这个城市也曾按同心圈层结构发展，但是由于地貌的不规则和西班牙殖民者不断把土地据为己有，其发展不久就呈现出线形的特征（图5）。

在17和18世纪，由于行政、商业以及住宅建筑增多，这个城市的中心区域得到了巩固并依照社会关系来布局，这种布局反映了经济、政治以及社会权利，

图4　海湾

图5　市中心的城市纹理

还有防御要求和占领者对劳动力的控制。在其宪法颁布之后，基多成为共和国的首都。20世纪初期，这个城市以及全国都开始了重要基础设施的建设，包括铁路系统的建造（1908年）连接了山区和沿海地区；有轨电车系统的建造（1914年）；水资源的更新；污水处理系统的建造和重铺城市道路。

可可制造的危机过后，主要出口产品和全国经济出现了第一次衰退，在这种情况下，大量移民从山区移到沿海，从农村移到城市。20世纪40年代，国家生产结构开始向城市地区集中。为了满足基本需要，社会压力在住房、服务、交通和城市空间这些方面的

转化中变得越来越突出。基多的扩张主要发生在两个方向：朝向城市中心的小山丘和朝向北部的平原。

20世纪50年代，经济的部分复苏和政治的稳定引起了广泛的建设活动。随后的半个世纪，基多城市的发展是以开发新城市区域为特征的，尤其是在北部的发展和城市的现代化（市场、道路、机场和公共建筑的建造）。非法居住点的快速增长和采用邻里单位组织街区来满足城市的需要同样成为城市的主要特征。

在20世纪60～70年代，城市转变的过程与农业土地结构和原油收益引起的资本主义现代化进程密切相关。由于农业用地所有权的不平等状态长期没有得到改革，从农村到城市的这种移民潮处于一种强烈的释放过程，经济危机和城市给予公共设施建设的补贴则强化了这一趋势。城市在增长，并形成了满足中等收入阶层需要的人居环境。国家空间-经济发展形成了基于基多和瓜亚基尔的双头结构，在运输中和基础设施服务的供应中形成了强烈的地区不均衡性。

在20世纪80～90年代，基多的发展是围绕着新的社会角色组织的，尤其是与金融业、商业中心和大规模的交通系统这些经济和服务活动相关。现在基多的发展是与聚集过程和一系列大规模基础设施建设及相关联的房地产投资活动，如城市南部的供水工程、在Tababela的Tumbaco峡谷建设的国际机场、通往机场的高速公路，拓宽城市向南及向北的道路，以及商业、金融和管理中心等各项服务设施的提供。由于整个基多峡谷地区的相对饱和状态，这个城市的扩张和增长要向邻近的峡谷发展。由于地理条件的限制，如陡坡、山地和生态保护区，扩张的过程阻止了城市空间肌理的连续发展。主要的扩展地区如下：向东部的Sangolqui、San Rafael、Conocoto、Amaguana、Cumbaya、Tumbaco；向北部的Pomasqui、San Antonio de Pichincha、Carcelen、Calderon和Carapungo；向南部的Guajalo和Guamani。

由于城市土地使用管理条例的限制，皮钦查山脉的西部形成了一个绿化环，东部则有一座城市公园（图6）。

四、全球化，机遇和挑战

基多和它的腹地有相当多的潜力，然而，它们也遇到了很多土地问题：

（1）城市空间和功能的结合很差，大都市周边地区与郊区的无序发展失控；

（2）城市中心区域的集中活动体现了对历史中心城市结构的过度使用，公路系统和基础设施呈现饱和状态；

（3）无节制地非法占用农业用地和环境保护区域经常导致环境灾难并且违反城市规范和现有土地使用法；

图6 从山上俯瞰海滨发展区

（4）投机过程导致城市肌理发展的不合理，以及出现相互脱节的状况；

（5）商业活动及居住区服务设施的建设积聚于主要发展走廊；

（6）工业中心分别集中于城市南部和北部，因此产生了完全不协调的各种活动，而且增加了劳动力的交通出行量；

（7）城市基础设施的短缺与结构和等级上的分布不均，导致了社会分层；

（8）公共交通系统的不足和不合理。

城市周边地区由于建造居民区而快速发展，这也催生了土地投机，引起城市结构的功能障碍和耕地占用，没有为未来留出足够的土地储备。

标准化却并不合适的空间模型在庇护关系型的城市管理实践下恶化。

五、城市战略

（一）基多战略规划 2020

1. 城市区域图景

根据这一战略规划，到 2020 年，基多作为一个安第斯（1969 年根据安地斯条约 Andean Pact 所成立的南美自由贸易组织，现有玻利维亚、哥伦比亚、厄瓜多尔、秘鲁和委内瑞拉 5 个会员国）大都市，将基于可持续发展的政策、原则与策略建成为一个拥有将近 330 万人口的现代化综合性城市，能够更加有效地发挥现代化大都市和区域中心的作用，并能够提升经济竞争力，促进可持续的经济社会发展，改善居民的生活质量。

此外，大都市区将融入整个国家体系以提升国家的重要性，比如：创造有序的城市空间，促进繁荣和团结，环境和美学品质良好，历史遗产区得到合理的保护和平等利用；预防自然和人为灾害；形成具有特点和民族自尊的团结社会并实现民主与管治。

2. 规划策略

（1）整合重建城市的宏观中心及其种种功能，如交通、运输、可达性、土地使用、建筑因素及基础设施；

（2）鼓励多中心发展，基于全区域的功能而建设整合的交通体系；

（3）通过整体与均衡的重建方式，使设施服务及城市系统得以完善，实现合理的城市发展；

（4）通过可持续的环境管理来保护自然和农业保护区并且减少城市地区的污染；

（5）实施全新的土地管理审议系统，这一分权管理的系统是基于集权化的系统而设立并且保障了对土地的有效管理；

（6）保护特色价值，重振历史区域活力，整合巩固城市肌理的增长，提升整个城市环境品质。

3. 规划结构

规划结构中考虑到以下主题而制定：

1）土地（城市化土地，适宜使用和不适宜使用的土地）

2）向心的系统

（1）通过保护和修复历史遗产区域和有争议地区，保护次要历史中心的形态和类型，改变城市形象和重建公共空间来实现城市中心区域的重建（图 7）。

（2）开发城市次中心以实现多中心均衡发展，为各区域提供同样的设施服务以减少社会居住分隔，同

图 7　圣弗朗西斯科教堂

时为老机场及其他城市边缘区提供一体化设施。

3）人口分布（到2020年整个都市区的人口合理数为317万）

4）道路和交通系统

（1）建立一个多方式和立体化的交通系统来保证合理的可达性和出行时间；

（2）沿着一系列不同的发展轴、公交站和相互贯通的地区间高速路体系来完善综合交通系统。

5）城市公共设施系统

目标是形成一个统一覆盖整个区域的设施系统，纠正现在主要设施集中于某些特定区域不合理的配置状况。

6）基础设施系统

关于城市重建的前提条件，人们已形成一个共识，要提高城市生产率和生活品质一定要处理好城市增长和基础设施服务的质量和覆盖率的平衡。

7）公共空间系统和城市形象的改进

根据其尺度、特点和用途的不同，在连续而不同层次的系统中整合公共空间，从而成为城市肌理和城镇形象的构建中的一个主要因素。

8）土地和房屋

通过形成多功能的城市肌理来实现居民的使用和服务，从而重构并优化城市增长方式。与之相随的是提供足够的设备和基础设施，适应居住建筑用地发展的需要。

9）管理系统

这个规划的实施需要重新调整城市管理系统以保证管理权限的落实和对土地的充分管理。为了达到这个目标需要三方面的保障：

（1）足够的机构编制；

（2）基于战略性规划和土地调整规划指导的机构系统规划；

（3）一系列的法律和金融工具（尤其是土地法规的实施）保证。

六、城市大规模开发项目

（一）土地调整规划（POT）

土地调整规划是战略规划中有关城市发展的经济、社会、文化以及生态政策的内容在空间物质性上的体现。土地调整规划中提出的城乡远景目标是通过充分规划的土地结构体系，寻找经济机会、平等的社会发展。基多市应该成为一个生产率高、环境可持续性强、生活质量得以持续提高的现代化大都市中心。

为了实现这些目标，将把交通系统引向沿海和境外市场（建造新的机场和道路系统）；建造货运枢纽和工业区，并改进服务，这些发展都将会使投资的产出更高；以基多为核心开展经济、法律、区域结构的现代化和多样化建设。

伴随而来的社会发展战略和项目将会提升一个平等、安全的城市，吸引投资并提升城市的竞争力。教育和职业培训项目应纳入规划考虑以增强竞争力，从而促进技术和应用科学的发展。

（二）新的都市圈用地规划

根据新的都市圈用地规划，最终将建立：

CBD（中央商务区）的新功能。它将保留政府机构、管理设施、金融和商务活动，但是那些在次级市中心能够得到发展的活动将不再集中于此，中心功能的疏解将改进交通可达性，提高道路设施的效率。此外，需要优先考虑的是纠正滥用土地的错误行为和促进城市更新的过程，这一点将基于对未开发城市空间的有效利用（图8）。

根据每个区域的特点，通过支持不同空间统一的物质性建设来进行新城市中心的整合。新中心的整

图8　海滨的栈桥区

合必须通过识别城市用地和城市潜在用地的"城市特性",重新定义城市界限的概念并利用城市的优势资源。城市生活空间的有效利用必须建立在2020年的预计人口规模基础上,依据不同活动发展空间的需要而规划。

对住房和经济活动的土地分配政策应该建立在城市发展的规律、城市化的新机制基础上,以减少各个层次上的无序,并满足当前和将来各层次人群的需要。

使用简洁的规则与标准,用灵活的管理来促使人居环境的整合,控制并加强城市结构以渐进式的方式实现城市化。

重组公共空间系统。通过一个整体的网络系统来组织空间层次与轴线,组织不同部门及其公共服务,以促进城市不同区域的和谐和功能性。

鼓励保留用地的开发。通过城市行政和服务项目的疏散,对社区设施网络系统进行永久性的改进以满足对当前和将来的土地和住房需求。

发展多方向多模式的流通系统（公路和交通）,这样可以把地区主要交通系统和新机场连接起来,扩大城市道路的交通容量,保证城市居民顺利出行。采用这种结构的话,重点将在公共交通系统上,但是同时也允许私家车在可持续增长的前提下适度发展（图9）。

将无轨电车系统作为空间、社会和环境重组的中心要素（图10）。

图9　海滨的高架路

图10　城市无轨电车枢纽

确定基础设施、设备和服务系统的要求。满足现在以及将来在容量、数量、质量和覆盖率方面的要求。

（三）城市形象的设计和改进

由土地调整规划（POT）提出了城市中心区的功能重整，并对城市一些区域进行了战略上和结构性的干涉：

北部的城市中心：Maridcal Sucre 机场周边地区城市化项目，Oyambaro 新机场的开放，以及它们与北部新城市中心的协调发展。

Villaflora 城市中心：城市南部的中心设计包括城市设施、基础管线以及城市活动的提供、整合与强化，例如无轨电车中转枢纽、El Camal 市场、El Recero 商业中心、la Villafloar 交通换乘中心和 Rodrigo de Chavez 大街等。

国家市民中心：长期以来公共机构遭受了很多有关滥用公共空间、私人小汽车交通膨胀和造成各种污染的指责，因此计划在 La Alameda、the Ejido、El Arbolito、立法宫及高等法院之间建设一个连续性的公园区以扭转公共机构每况愈下的形象。

对 Alameda Axe 大街、Pichincha 大街及 Machángara 城市区域实施修复：恢复道路通行能力，改善城市中心和城市南部的联系；修复 Machangara 和 Grande 河盆地地区的环境，为城市南部的居民提供休闲娱乐场所。

历史街区的重建：调整历史地区的总体规划，尤其是在城市中心的区域。这将考虑到结构性项目的开发，比如：非正式商业活动的合理化、交通流动的合理化、改造低收入区域（La Mariscal、Larrea、Santa Clara、Belisario、Chimbacalle-La Recoleta、El Placer-San Roque、La Chilena-El Tejar），以及对重要建筑的修复（包括原 García Moreno 监狱和军人医院等）。

改进城市形象的规划：改进和恢复公共空间和城市形象，要把设计和重新设计城市各区域的可识别性作为市政府的一项长期性任务。初期项目是在 Veintimilla 街、Naciones Unidas 大街在 Michelena 区域的部分、厄瓜多尔中心大学和基多天主教大学之间的区域完善人行设施系统，另一个项目是改善厄瓜多尔中心大学校园形象的。有一些项目通过"IDEAS PARA QUITO 基多点子竞赛"的形式进行，吸引了包括建筑师和城市规划师在内广大市民人群的参与。

（四）其他的不同项目

1. 经济发展

Itulcachi 工业园区项目：已经有 30 多个有危害性的工厂入驻进来，这些工厂如不加合理安置将对环境造成很大污染。

城市南部流行商业中心：历史中心公共空间得到重建后大量非正式的商业活动涌入，传统的 Camal de Quito 地区就形成了 1200 个工作岗位。

公墓：现有的墓园不能满足当前的需求，尤其那些低收入的人群。市政当局开始建设两个新公墓，一个在东南部（占地 $9.3hm^2$，有 24400 个墓穴），一个在北部（占地 $11.4hm^2$，有 30000 个墓穴）。

公园：在 Bellavista 山旁的基多市公园（大约 $650hm^2$），面向城市的东北部；Itchimbia 地区公园（大约 $53hm^2$），位于 Itchimabia 山下。

最后确认了一些当前需要进行的规划基础工作，它们是：

（1）研究：大都市地区的土地制度、房屋政策和人居环境，基多的土地和土地市场，基多的城市环境，基多的教区和邻近区域的 Mosaic 全球定位系统，基多的郊区原型研究。

（2）大都市信息系统：城市数据维护、科技信息支持、大城市信息库、基础地图、互联网服务、流行病预警系统、在邻近地区的城市投资、机构协作。

（3）灾难防护：城市建筑的功能，有意识的防护活动和紧急事件应对程序。报告的详细内容和建议包括：加强学校建筑抗震能力的工程；厄瓜多尔建筑规范；基多市的地震预防；在基多都市区非稳定土质的探测项目；召开区域会议，组织针对安第斯地区和南部火山区域的灾难防范应对活动；成立灾害紧急事件的应对委员会（图11）。

(4) 历史区域的修复：为特殊的历史街区制定区划，界定能够新建建筑的地区；制定受保护建筑和郊区教堂的详细清单；调整 La Mariscal 精选目录；增录基多历史中心的名单；确定 Rumicucho 考古发掘区的土地利用和土地出让条例。

(5) 保护 Camino del Inca 的法令

图11　受到保护的城市历史区域

七、结论

"厄瓜多尔"在西班牙文里就是"赤道"的意思，首都基多，则是从古代居住在这里的古印第安人"基图贝"部落的名字演绎而来的。七八百年前，印第安人就在这里建立了基多王国，15世纪末，基多实际上成为印加帝国的第二首都，是著名的印加文明的中心之一。尽管经过1917年的地震，基多仍然是拉丁美洲保存最好、改变最小的历史中心。圣弗朗西斯修道院和圣多明哥修道院，拉孔帕尼亚的教堂和耶稣会学院，连同这些建筑华丽的内部装饰都成为了"基多巴洛克风格"的纯正典范，完美地融合了意大利、摩尔、佛兰芒和当地艺术的精华。如今，它已被联合国教育、科学及文化组织列入世界文化与自然遗产保护名录。基多作为"赤道之国"的首都，依然闪烁着赤道文明的光辉。

面对全球化的挑战，基多将基于可持续发展的政策、原则与策略，建成为一个拥有将近330万人口的现代化综合性城市，能够更加有效地发挥现代化大都市和区域中心的作用，并能够提升经济竞争力，促进可持续的经济社会发展，改善居民的生活质量。

参考书目

[1] Plan General de Desarrollo Territorial del Distrito Metropolitano de Quito [Z]. Dirección General de Planificación del Municipio del Distrito Metropolitano de Quito, 2001.

[2] Quito Regimen Distrital del Suelo, Propuesta [Z], op. cit.

[3] Barrera Augusto G (coordonator). Ecuador: un modelo para (des) armar. Descentralización Disparidades Regionales y Modo de Desarrollo [Z]. Producciones Digitales UPS, Ecuador, 1999.

[4] MDMQ – DGP. La Transición a Distrito Metropolitano de Quito, Quito [Z], 1995.

[5] MDMQ – DGP. A-Régimen Distrital de Suelos, Propuesta, Quito [Z], 1996.

[6] MDMQ – DGP. B-El Mercado de Suelo en Quito, Quito [Z], 1996.

[7] MDMQ – DGP. Plan General de Desarrollo Territorial, Quito [Z], 2001.

[8] MDMQ：Municipio del Distrito Metropolitano de Quito [Z].

[9] DGP：Dirección General de Planificación [Z].

[10] LRM：Ley del Régimen Municipal [Z].

[11] LDMQ：Ley del Distrito Metropolitano de Quito [Z].

QUITO

Diego Carrión

Located in Andean valley running from north to south at a height of 2820m, Quito lies between the Pichincha volcano in the west and the Itchimbia hills in the East. The urban structure has been conditioned by the scarcity of flat land, the topographic irregularities of the surrounding mountain system and the numerous east-west slopes.

Since 1993 the Law of the Metropolitan District of Quito (LDMQ) has provided a wider legal framework than the traditional municipal competencies outlined in the Municipal Law. The LDMQ is a legal instrument only applicable to the urban and rural management of the city of Quito. This Law has generated important administrative changes in the metropolitan area, with respect to the decentralisation of the city's management.

During the 17th and 18th centuries the urban area of the city was consolidated with administrative, commercials and residential buildings, organised according to social relations, which expressed economic, ideological political and social hegemony, defence requirements and the availability of labour for the conquerors.

After the proclamation of its constitution Quito became the capital of the Republic. In the early 20th century important infrastructure works were carried out in the city and in the country. These included: the construction of the railway system (1908), which linked the Sierra with the Coast (Quito with Guayaquil); the construction of the tram system (1914); the renewal of the water and sewage system (1922) and the re-paving of the city.

After the crisis in cocoa production, the first export-staple and a national economic recession, there was substantial migration from the Sierra to the Coast and from the rural to urban areas. In the 40s, the transformation of the national territorial productive structure towards urban areas started. The social pressure for the satisfaction of basic needs such as housing, services, infrastructure, transport and urban spaces became important in the shaping of these transformations. The expansion of Quito took place in two directions: towards the hills in the centre of the city and towards the Iñaquito plain in the North.

In the 50s a partial economic recovery and political stability gave rise to extensive construction activities. The urban development of Quito in the last half of the century is characterised by the opening up of new urban areas, especially towards the North and by the "modernisation" of the city (construction of markets, roads, the airport and public buildings). The rapid growth of illegal settlements and the use of neighbourhood organisations to make urban demands also became a major feature.

During the 60s and 70s the urban transformation process became linked to the capitalist modernisation of the agrarian structure and the availability of oil revenues. A strong rural-urban migratory process was unleashed associated with the lack of reform of the unequal land ownership system in agriculture; the economic crisis and city subsidies for the construction of public works. The cities grew and settlements were formed to house medium income sectors, the sub-proletariat and, in Quito, the bureaucracy. The national space-economy developed a bicephalic structure based on Quito and Guayaquil, with strong regional inequalities in the delivery and the supply of infrastructure and services.

In the 80s and 90s, the city of Quito was organised around new social actors, particularly those linked to finances and to service activities as commercial centres and mass transport systems were developed. The growth of Quito is currently linked to a process of agglomeration and real estate speculation focused on large infrastructure works: projects for the supply of water towards the South of the city (Project Mica -Quito Sur), the construction of the international airport in the Tumbaco valley in the Tababela sector, the construction of the highway to the airport, the enlargement and improvement of access roads to the North and South of the city, and the supply of services (commercial, financial and administrative centres).

Due to the relative saturation of the consolidated area in the Quito Valley, the expansion and growth of the city tends to be oriented towards the neighbouring valleys. This process hinders the continuous development of the urban fabric due to the geographic accidents such as steep slopes and hillsides and zones of ecological protection.

A 'green ring' has formed in the west along the Pichincha mountains given the existence of restrictive municipal land use regulations, and in the east by the presence of the metropolitan Park.

Globalisation and Strategy

For the year 2020, the Metropolitan District of Quito will be a modern territorial and urban complex with approximately 3.3 million inhabitants, consolidated as an Andean Metroplolis and developed under sustainable principles, policies and strategies.

The Metrolplitan District of Quito will be integrated to the country strengthening national values such as an ordered space, welfare and solidarity, environmental and aesthetic qualities, its historic and heritage areas protected, preserved and used with equity; safety upon natural and human risks; governable and democratic for a solidarious society with identity and self-respect.

The Strategies of the Plan

● The integrated restructuring of the macro-centrality of the city and its functions in relation to traffic, transport, accessibility, uses, building factors, services and infrastructure.

● The encouragement of a poly-nodal system of centralities and the development of an integrated poly-modal system of mobility and accessibility based on the functionality of the whole territory.

● The rationalisation of growth, densification and urban development, through the restructuring of services and the urban system in an integrated and balanced way.

● The sustainable management of the environment to preserve the natural and agrarian reserve areas and to reduce the high level of contamination of urban areas.

● The implementation of a new judicial system of territorial management that is decentralised and based on the system of centralities and which assures the management of land.

● The protection of identity values, the recuperation and revitalisation of historical areas, the consolidation and growth of the urban tissue and a generalised improvement of the environmental quality of the whole territory.

The Structure of the Plan

The Plan is structured in relation to the System of Centralities:

● restructuring of the central metropolitan area through the implementation of rehabilitation programs in the zones of historic patrimony and conflict zones; through the protection of the morphology and typology of secondary historic centres; through improvement to the urban image and the restructuring the public space;

● restructuring the metropolitan sub-centres towards a de-concentrated and harmonious development that softens the degree of urban-residential segregation through the equitable supply of collective equipment in a set of special projects (old airport, Ciudad Quitumbe) and through projects for the integrated development of the various urban and peri-urban zones.

1) Road and Transport System
2) Urban Equipment System
3) Basic Infrastructure System
4) Public Space System and Improvement of the Urban Image
5) Land and Housing
6) Management System

Strategic Projects

The plan of Territorial Ordering (POT), constitutes the physical expression of the economic, social, cultural and ecological policy of urban development established parallel to the Strategic Plan. The Urban-Regional vision proposed by the POT considers Quito should become a modern metropolitan centre that can achieve urban productivity, environmental sustainability and improvements to the quality of life through an adequate territorial structure, and the search for economic opportunities and equitable social development.

These goals will be achieved: through the orientation of the communications systems towards the coast and external markets (the construction of the new airport and road system); through the development of freight terminals and industrial zones and services improvements which will support productive investments and through the modernisation and the diversification of the economic, legal and regional structure that has Quito as its core.

The accompanying Social Development strategies and programs will help to promote an equitable and safe city, and to attract the investments that will enhance the competitiveness of the city. Education and training programs should be designed to enhance competitiveness and to promote the development of technology and the applied sciences.

The New Territorial Metropolitan Structure

It will establish a new functionality for the CBD and retain government, administrative equipment, financial, professional and commercial activities, but those activities that are able to be developed in sub-urban centres will be deconcentrated, thus improving accessibility, and increasing road infrastructure efficiency.

The integration of new centralities in the city through supporting integrated physical development solutions at the different spatial levels, in accordance with the particularities of each zone. The integration of new centralities should be made by recognising the "urban" character of both the urbanised and the potentially urbanisable land; by redefining the juridical concept of the urban boundary and by taking advantage of land resources.

Land delivery policies for housing and economic activities. Should be based on urban principles, new mechanism of urbanisation that reduce the levels of informality and satisfaction of the current and future needs of all segments of the population.

To promote and to elaborate clear rules and simple norms. For the control and strengthening of progressive urbanisation of the urban structure with a flexible management that facilitates the consolidation of human settlements.

To restructure the public space system. In a way that integrates and articulates the different sectors and public equipment through a consolidated network and which defines axes and hierarchically - organised spaces that foster the harmony and functionality of the different areas and zones of the city.

To encourage the creation of land reserves. For municipal projects for administrative and services decentralisation, the permanent improvement of the community equipment network and the satisfaction of the current and future demand of land and housing.

To generate a multidirectional and multimodal system of mobility (road and transport). That can articulate the primary system with regional access and the new airport, and which is able to adjust the road capacity of the city, to guarantee accessibility and support for the city's residents. Within this structure, the emphasis will be on public transport, but with allowance for the sustainable growth of private cars.

To determine the requirements for infrastructure, equipment and services system. To meet present and future demand in terms of capacity, quantity, quality and coverage.

Design and Improvement of the Urban Image

The refunctionalisation of the central area, proposed by the POT establishes a strategic and structured intervention upon certain parts or territorial components of urban space:

● North Urban Centre: The rational and programmed urbanisation of the Mariscal Sucre Airport and the opening of the new airport in Oyambaro, and its harmonisation with the new urban centre in the North Zone.

● Villaflora Urban Centre: The design of an urban centre in the Southern Zone of the City, based on the provision, agglomeration and renewal of equipment, infrastructures and urban events, such as the trolley-bus transfer station, the El Camal market, the El Recero commercial centre, the la Villaflora traffic transfer and the Av. Rodrigo de Chávez.

● National Civic Centre: The formation of a park capable of articulating and redefining the continuous public character of the space between La Alameda, the Ejido, El Arbolito, the Legislative Palace and the Supreme Court. Current downgrading trends will be reversed and its symbolism restored, having been affected by the use and abuse of public space, the huge volumes of private car traffic and various forms of pollution.

● The urban rehabilitation of the Alameda Axe - Av. Pichincha - Av. Machángara sector through: the recovery of road capacity, the improvement of the connection between the centre and the south of the city; the environmental recovery of the Machángara and Grande river basins and finally, the supply of recreational areas for the population in the southern part of the city.

- Rehabilitation of Historic Areas. The readjustment to the Master Plan of Historical Areas of the District, particularly those located in the city centre. This will consider the development of structural projects, such as: the rationalisation of informal commercial activities, the rationalisation of traffic flows; the upgrading of low income 'barrios' (La Mariscal, Larrea, Santa Clara, Belisario, Chimbacalle-La Recoleta, El Placer-San Roque, La Chilena-El Tejar); the architectural rehabilitation of important buildings (the former García Moreno jail and the Militar Hospital, etc)

- Plan for the Improvement of the Urban Image: The enhancement and recuperation of the public space and the urban image of the city through design or redesign of parts of the city identified as pilot models for what will become a permanent task of the municipality. Preliminary projects will be oriented towards creating pedestrian facilities in the street Veintimilla, av. Naciones Unidas, in the Michelena sector and in the sectors between the Central University of Ecuador and the Catholic University of Quito. There is also a project to improve the urban image of the University Campus of the Central University of Ecuador. Some of these projects will be implemented through public competition in the programme "IDEAS PARA QUITO" that includes the widespread participation of the citizenry, architects and urban planners.

里约热内卢（巴西）

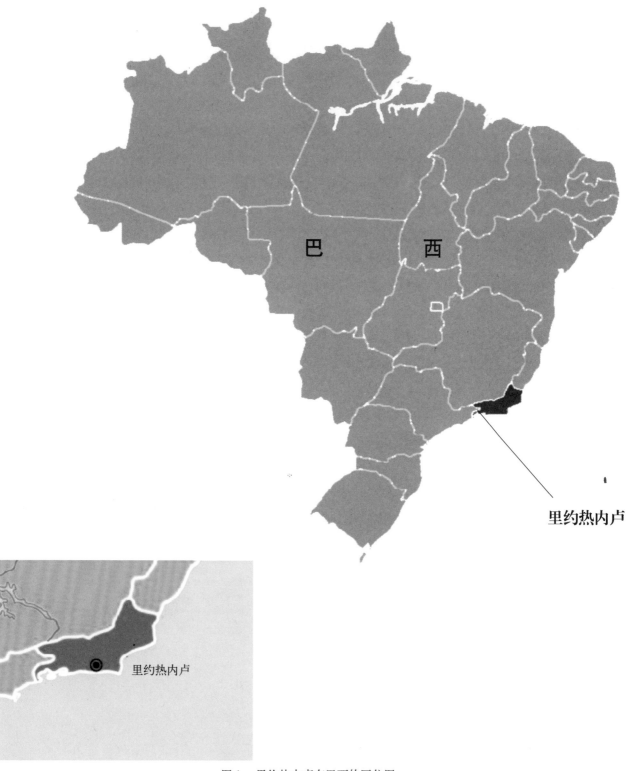

图1　里约热内卢在巴西的区位图

案 例

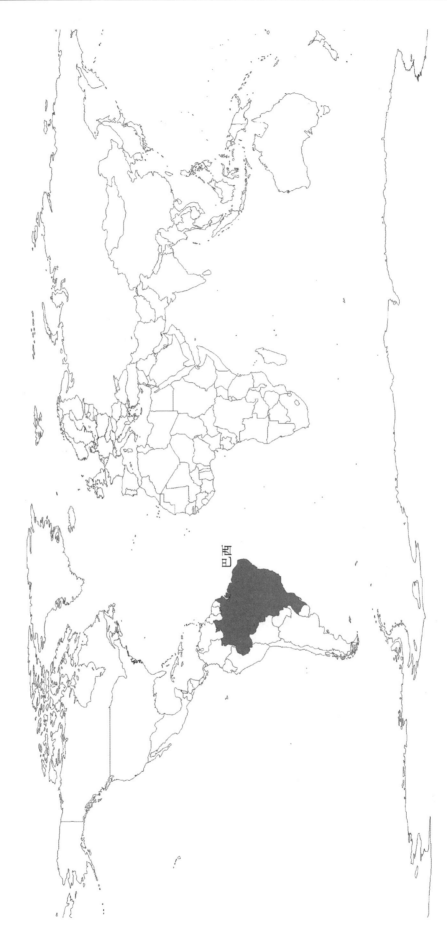

图 2　巴西的全球位置图

里约热内卢

佩德罗 · 约根森

一、城市数据

（一）人口和用地

面积：1264.2km²。

大都市圈（包括19个自治市）：5693.5km²。

人口（2000年）：

里约热内卢州人口：14391262人。

大都市圈内人口：10872768人。

市区人口：5857904人。

年增长率（1991～2000年）：

巴西全国（1991～1996年）：1.36%。

里约热内卢州：1.28%。

大都市圈：1.14%。

市区：0.73%。

1997年人均总收入与总支出（雷亚尔）：

其他州府：513536（约228275美元）。

里约热内卢：536590。

1998年国民生产总值：

里约热内卢州：100616百万美元。

直辖市：60578百万美元。

（二）经济指标

1. 劳动力状况

经济活动人口（PEA）：

大都市圈：435.3219万。

市区：230.9390万。

失业率：9%。

生产部门中的就业人口：21%。

正式部门中的就业人口：52%。

收入少于2种最低工资标准的就业人口：31%。

（三）住房、城市基础建设和其他社会指标

1. 基础设施覆盖率

给水：95%。

系统覆盖区域：69%。

系统外的区域：5%。

住房（大都市圈——2000年）：

住房总数量：2623342套。

面积不足数量（住房短缺和共居）：286951套。

条件不足数量（基础设施短缺和密度过高）：1218367套。

（四）教育，健康和休闲

1. 健康状况

联邦健康体系：89所医院、15000个床位。

基本数据：

儿童死亡率：20‰。

期望寿命：65岁。

2. 教育状况

15岁以上文盲率：4.4%。

15岁以上受过学校教育少于4年：12.9%。

特殊教育：国家聋哑教育中心、本杰明教育中心（视力残障）、残障军人中心、军队培训学校、高级战争学校。

高级知识中心：奥斯瓦尔多·克鲁兹基金会（医疗和生物）、里约热内卢大学研究中心（政治科学和社会学）、杰杜里奥·瓦卡斯基金会（经济和行政管理）、理论数学和应用数学中心、天主教大学的情报学实验室、巴西物理研究中心。

3. 交通运输

公交车：占日运输总量的77%。

小汽车和出租车：占日运输总量的14%。

地铁（私有）：40万人次/天。

市郊火车（私有）：24万人次/天。

船只（从里约市往尼特罗伊港、巴克达岛和格布纳多岛）：占日运输总量的1%。

国际航空：24个国内站点和85个国际站点。

国内航空：每30min开往巴西的各主要城市。

里约热内卢港：拥有石油和石油产品终点站（吞吐量达每年1400万t）和一个旅游终点站。

塞巴蒂巴港（位于伊塔瓜伊自治市）吞吐能力达15万t。

4. 文化

4份全国发行的报纸。

1个全国电视网。

92座博物馆和文化中心。

巴西交响乐团、市立舞蹈剧场公司。

公众节庆活动：狂欢节和新年。

5. 旅游业

外国游客（1999年）：172.4万人。

旅馆设施：188个旅馆，容量约2.3万人。

二、城市概况

经济衰退、沉重的债务以及1998年针对宪法改革的民主化进程，为里约热内卢的地方区域自治创造了条件。这些社会变革促使政府重新审视土地及基础设施发展规划，以及一直推行的根除贫民区的政策，贫民区的更新和城市中心区的保护开始引起重视。1992年颁布的十年计划旨在推进基于社会公正的城市民主化管理。20世纪90年代，相关法律机制也在城市管理和贫民区更新等领域开始出现，然而新增土地的管理并未规范化。

20世纪90年代的民主化进程与经济增长放缓、经济领域里不公平的增加、非正规经济的扩大等各类负面情况同时出现。现行城市发展战略将大规模公共投资（Favela-bairro，Rio-Cidade）与城市设施的特许经营项目（Linha Amarela，mobiliario urbano）结合在一起，用于面向世界市场建设具有竞争力的综合城市，此计划将通过大力发展旅游业以及通信枢纽建设、滨水区建设、2004年奥运会等大型活动庆典来实现。新的千年伊始，这些大型活动包括庆祝港口地区的复兴、2007年泛美运动会以及古根海姆博物馆的建设。

三、城市历史

圣塞瓦斯蒂安·里约热内卢建于1567年，建城之初是作为控制瓜纳巴拉海湾的一个要塞。18世纪，吉赫金矿成为了葡萄牙王国的主要财富来源，使殖民路线延伸到西南部。在1763年，奴隶、小商人、军事和政界的精英在这里定居。

1808年，受到拿破仑入侵的威胁，殖民地首府（大约1.5万人）从巴西的西南迁移到大咖啡生产区的经济中心。港口的开放打破了殖民地的统治，城市经历了巨大变化。1822年，巴西独立，里约成为巴西帝国的首都。独立战争结束后，由于人口增加、国外投资涌入和运输方式的机械化，里约开始向现代化城市发展的步伐。

1835年，船舶供给服务从市中心转移到了卡居港（Botafogo y São Cristobão）。在19世纪40年代的10

年间，新的船运航线把市中心与尼特罗伊海湾的另一侧联系在了一起。从 1850 年开始，政府把公共设施经营权承包给了外国公司。1854 年，公共煤气开始使用，1862 年，公共卫生设施也交给外国公司经营。在此期间，大量的人口仍然居住在市中心，他们依赖于优越的地理位置以谋生，因此，尽管有了这些新的服务设施，城市还是遭遇了黄热病的流行。

铁路的发展推动了市郊的形成，并逐步突破城市的边界形成大都市的空间格局。1868 年，原来靠畜力牵引的轨道车将交通服务延伸到城市新区，如沿海地区和山区。有轨电车的经营者们在土地拥有者的配合下完成了大型基础设施工程，如 1892 年和 1890 年建成通往克帕卡巴那（Copacabana）的两条隧道。为推动公共设施建设，市政当局明确了公共设施的所有权。

1900 年，里约市作为一个农业共和国（1889 年巴西成立共和国）的首都已有 50 万居民，且巩固了从事咖啡生产的中产阶级的统治地位。现代化发展的需要使城市采取了两方面的行动，首先是建设城市公共卫生系统和根除传染病，其次是从皮耶拉·帕索市长开始的奥斯曼式的大规模城市改建。这些行动的结果是，为建设中央大道和门德萨大街，由于奴隶解放（1888 年）产生的大量贫困人群从市中心迁出，并在 1910 年建成了新的港口。

在 20 世纪 20 年代期间，摩洛多卡斯特罗山（Morro do Castelo）被夷为平地，法国的城市规划专家阿尔弗雷德·阿加西（Alfred Agache）将美术和城市美化的理论与城市工程建设的要素相结合，设计了"改造与美化"方案。该方案的设计要素在瓦加斯统治时期（1930～1945 年）仍然得到推行。工业化的深入、工人阶级和城市中产阶级的形成，以及劳资关系法案的出台，奠定了巴西现代化的基础。在 1940 年，巴西大道建成开放，它成为了通往城市的重要道路，同时启用的还有瓦加斯总统大道，这是一条 100m 宽的大道，由两侧林立的高楼构成了纪念性的轴线，该项目通过出售路旁土地获得的资金得以建成。

从 1940～1960 年，里约市已经发展成为大都市。国家政府大力支持工业化，使得农村劳动力转移到缺乏公共服务设施的城市外围地区。城区人口从 180 万增至 330 万，而整个城市人口从 220 万增至 490 万，城市问题日益严重，如贫民窟的数量增加，交通拥堵状况加剧，可饮用水缺乏。轿车工业成为经济发展的龙头和现代化的标志。城市的管理此时主要集中在道路建设和贫民窟的清除上。道萨迪亚斯（Doxiadis）的规划创造了多样化的道路。公共汽车和小汽车代替了有轨电车，同时 Aterro do Flamengo 专用车道、Praça XV 高架高速路、长距离隧道和城市中心区外围的大型住宅区也相继落成（图 3）。

1960 年首都迁移到巴西利亚并未影响里约热内卢作为该国最重要城市的地位。由于拥有重要港口、遗留的联邦公务人员、工业与服务园区，里约热内卢保持了一个大都市的重要性。20 世纪 70 年代，中央集权专政者设立了大都市区，制定了城市规划的程序，增建了许多大型的公共设施，如尼特罗伊大桥和地铁。1975 年，里约热内卢成为新成立的里约热内卢州首府。城市基本规划（PUB-RIO）是一个将

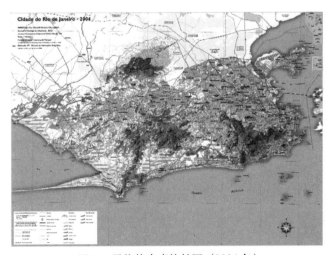

图 3　里约热内卢航拍图（2004 年）
（来源：里约热内卢市文化遗产事务局）

图 4　港区边缘的 Francisco Bicalho 大道
（来源：P. Jorgenssen Jr.）

综合交通体系规划以及通过建设主城区外围大型购物中心与机动化交通来促进 Baixada de Jacarepaguá 发展的科斯塔（Lucio Costa）计划大大推动了城市向外发展（图4）。

四、全球化，机遇和挑战

里约热内卢一直在全球商品和服务市场上具有综合的竞争力，同时也具有建设大规模城市发展项目的悠久历史，然而，至今还没有一个面向全球化时代的城市发展规划。

将里约热内卢建设成为大型枢纽港的计划始于20世纪80年代，1993年召开了"里约—巴塞罗那城市发展战略"研讨会以后，这项计划正式成为"城市管理文化"的一个组成部分，从此，城市以提升竞争优势为目标，邀请大量的国际咨询机构参与了规划、研究及项目建设工作。这部分的发展历程将在后文的"战略规划及城市大规模开发项目"部分集中阐述，以港口地区复兴和古根海姆博物馆建设这两个最重要的项目作为实例。

因经济壁垒而产生的全球化、市场开放的不对称和通信科技的发展带来了城市管理思路在20世纪80年代的变化。这段时期的两个最主要特征分别是产业结构由制造业转向服务产业，以及边缘化国家[①]的债务危机造成公共投资锐减，这影响到大型公共项目的效益，特别是影响了1950～1980年间建成的基础设施项目发挥作用。

随着城市发展的目标调整为建设"具有竞争力的城市"，以吸引跨国企业的投资（制造业及金融业）、举办大型庆典活动以及发展旅游。城市中心区聚集了房地产及其相关产业，因为这些产业的高额利润足以支付城市基础设施建设和城市化的成本。

通过把城市旧区更新改造成为现代都市区成了城市管理的最佳模式，与此同时，城市中心的建筑、历史和文化遗产保护的问题再度成为争论的话题，这些倾向在20世纪70年代末的巴西显现，最初是对纯粹的功能分区为主导的规划是一种批判，在80年代末则成为规划政策的补充。

新的城市管理模式来源于全球化并且服务于全球化，包括把荒废的工业区、码头、铁路改造成新的商业中心、工业园区、历史文化旅游区和中产阶级居住区，通常这些项目位于城市的中心区及其周围，有时候则位于重要的交通枢纽。

在经济低增长和低负债的国家，这些项目用来减轻公共与私人投资不足所带来的诸如失业、经济非正规化、不平等、贫困等种种问题造成的负面影响。由于经济全球化本身会导致发展的不均衡，因此对于面临城市退化的城市规划师和面临经济不景气的经济学家来说，大规模的城市项目建设成了城市更新的当务之急，也是缓解经济衰退的途径。

目前，大规模城市开发项目主要有以下四个特征：

（1）大规模城市开发项目不再只是基础设施建设、城市化或者房屋的建设，而是整个城市区域的整合重组或重建，以此创建吸引人的城市环境，实现销售和消费活动；

---

① 边缘化国家 Peripheral Countries，相对于全球发达城市而言。

（2）部分公共投资是以非货币形式投入的，比如说在原有荒废的公共事业用地上加以建设。现有的研究并未表明这些非货币形式的投入是否会造成新的私有产品减少；

（3）在私人项目的经营管理框架下，稳定地价的举措（法规和有利的外部效应）控制所产生的结余可以用来支付基础设施和公共事业建设的费用；

（4）通过稳定地价的措施（如法规和有利的外部配套），可以规范与管理私人开发项目，并获得基础设施和公共事业建设需要的资金。

我们所面临的是前所未有的挑战，要管理这样的项目建设需要全面掌握从设计技术到城市、环境与行政法规，以及城市管理、交通、社会、经济与方案评估等多方面的知识，也需要有效而透明化的公共事务管理机制。另一方面，经济全球化对边缘化国家的冲击之一就是垂直化与公共事务管理能力的减弱。大规模城市建设项目的谈判与决策制定过程涉及的范围越来越小，离开公众监督也越来越远。

由于公共空间价值的再发现与一些成功案例的经验传播，投资者热情高涨，城市规划决策者们也认为当前是城市发展的黄金时期，然而试图通过面向成为全球经济、旅游及重大事件中心的城市发展战略能够对边缘化国家的大城市产生重要而持续的积极影响几乎是不大可能的。

除了对基本而迫切的住房问题和贫民区改善问题——也就是通常所说的城市改造——缺乏足够的反应以外，大规模建设项目也没有对其他的城市系统性问题发挥作用，如城市品质、效率、标准化、为公众提供服务设施、项目与系统管理的透明化、城市疏解以及在重大决策的制定上公众的有效参与。

五、战略规划

到 2003 年，里约热内卢的战略规划已实施了十年。这个过程以"里约—巴塞罗那城市战略"研讨会为起点，在 1995～1997 年间，该战略规划由来自西班牙巴塞罗那以及巴西的城市和商业领域的专家进行详细的研究。十年指导性规划则是由市议会批准后在 1992～2002 年期间实施，根据 1998 年的国家宪法，城市指导性规划具有法律效力，其制定的原则是通过对社会财富的再分配与民主价值的形成来促进城市管理。因此，尽管战略规划与指导性规划被期望弥补城市管理在规范上和操作上的不足，但两者自身仍然包含了相互矛盾的观点，比如公共政策的目标、执行机构、实现的方法等。

在此期间，城市发展战略得到有效的贯彻实施，尽管项目最初的原则和指导方针框架并非由战略规划决定。这些目标可以归纳为以下四点：

城市发展目标主要是增强城市在咨询、工程建设、大型庆典和城市服务设施等方面的全球综合竞争力。例如枢纽港建设、黄线高速路（Linha Amarela）建设经营权的转让、申办 2004 年奥运会、战略发展计划、城市街道设施特许经营的国际招标、内城复兴标志项目的招标、港口复兴、古根海姆博物馆的兴建和 2007 年的泛美运动会。

制度化与规范化的弹性：由州立机构决定选择战略上的规划或者代表民意的市议会提出的建议，有投资和提议能力的社会机构也可以参政，如它们解散了现存的城市政策委员会，多次使用"联动运作"的方式。

公众投资积极促进了社会经济和文化方面的发展：Favela-Bairro 和 Rio-Cidade 计划加强了个体经济与小型企业的活力，从而达到提升城市形象的目标。这些投资的效果尚未完全显现，但已经可以看到贫民区居民住房条件的改善与生活质量的提升，以及相关商业街区持久性经济行为的增长，现在这些项目不仅国内闻名，还名扬海外（图5、图6）。

通过"虚拟项目"营销城市：一些项目看上去并不关注公众的基本需求，而是为了进一步吸引投资。项目的进行未能事先获得政府批准，并缺乏可行性研

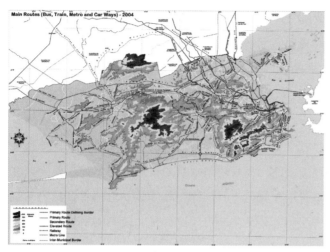

图5 主要道路体系
（来源：里约热内卢市文化遗产事务局）

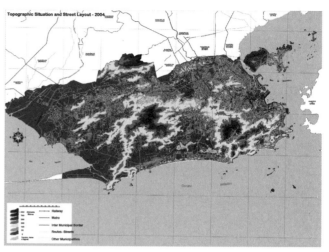

图6 城市地形与道路格局
（来源：里约热内卢市文化遗产事务局）

究、资源使用及管理架构。因此，项目的建筑效果图在此起到项目宣传推广和可行性测试的双重作用，好像它们建成后能够像市中心的那些高档建筑一样。事实上，一些可行的建设项目倒不需要详细的可行性计划作为保障。

在2001～2004年期间，战略规划的焦点发生了重要的转变，转向着眼内在经济的发展、地方科学技术机构的参与以及网络建设。新的规划组成包括：①区域战略规划；②可参与的预算；③全球规划的实现；④对不同项目的改进、监控和效果评价。

六、城市大规模开发项目

（一）通信港

通信港项目的含义就是用最新通信技术装备的商务活动区。这个概念1993年首次提出，后来包括在Estácio-Cidade Nova 地区的重建计划中。在25万 m² 的土地上，已建成6.2万 m²，待建面积达45万 m²，这一地区包括了主要的市政中心以及一个地铁站点。它同时也紧邻金融中心、机场（包括一个国际机场和一个国内机场）以及港口。

- 市政当局投资3100万雷亚尔（约合1378万美元）建设城市基础设施，两栋私有建筑已建成，网络接入点和可容纳3000人的会议中心已经列入计划。总投资已经达到了1亿3000万美元，在接下来的三年里将再投资2亿6000万美元。

（二）第十五广场（Praça XV de Novembro）重建工程

这个广场坐落在瓜纳巴拉海湾的滨水区，它是包括了1600座修复建筑的历史中心。重建工程包括拆除高架步行道、建设地下通道、重新设计公共空间，以及恢复通向PraÇa Imperial（帝王广场）和Chafariz de Mestre Valentim 的通道，其结果是增加了各种都市活动与步行的频率，特别是在周末。1988年，二期项目将约1km长的滨水空间改建成为公共活动与文化活动的场所。

（三）黄线高速路工程（Linha Amarela）

一条25km长的高速公路把国际机场和高速路与Barra da Tijuca区连接了起来（该区集中了高收入人群）。这是巴西第一条收费高速路，在政府的特许下该公路主要由私人投资。这项工程除了因为私人交通具有独占性而带来问题外,由于周边被征用的地块迅速被临时定居点所占据,该区域的城市复兴项目没有得到顺利实施。

(四)贫民区改造计划

这项计划的目的在于通过政府补贴的投资和居民自身的努力,来改善居民生活,提供生活必需的排水设施、公共服务,同时,通过维持地区秩序、职业培训与谋生技能培训,将贫民区整合到邻近的社会和城市体系中去。

这个计划的投资来自于城市财政投资,以及联邦和美洲开发银行(IDB)提供的贷款,此外还包括由欧盟提供的无偿援助。项目涉及 104 个社区,其中 28 个小型社区(规模少于 500 户,共计 4.3 万人)、73 个中型社区(规模 500~2500 户,共计 25 万人)和 3 个大型社区(规模超过 2500 户,共 8.3 万人)。第二阶段的改造计划包括 63 个贫民区,涉及 30 万居民,已确定获得美洲开发银行提供 3 亿美元的财务支持。

(五)里约-斯大德计划(Rio-Cidade)

这项计划旨在使逐步恶化的公共空间逐步恢复生机,这些公共空间由于 20 世纪 50~60 年代的大规模道路基础设施建设和 80~90 年代的购物中心以及高档社区的建设而被破坏,这一策略旨在通过具有象征性和集中的行动产生事半功倍的效果。先期选出的 17 条商业街都进行了公开投标,投标书包括开发方式、设计方案及建设标准。现在,许多项目已建成,有些则在与受影响社区进行磋商,或者与市政当局进行技术谈判。

(六)码头区的复兴规划——里约港/古根海姆博物馆

码头区的复兴计划和古根海姆博物馆的建设大半是独立投资的,不足部分来自政府拨款,博物馆选址

图 7 地铁建成后的第十五广场,改建项目由建筑师 Oriol Bohigas 负责
(来源:P. Jorgenssen Jr.)

在 Maua 广场码头处,位于码头区与 Rio Brance 大街(城市金融中心)的交汇处,是整个码头区域最重要的地块,其开发将是整个里约热内卢最宏伟的项目之一。

政府所属的城市研究所负责制定整个港区的发展规划。尽管没有形成清晰的管理模式——这是项目成功的重要前提之一,很多准备工作已经开始,如规划设计、工程及配套公共设施的合同谈判等。

这两个开发项目的结合是长达十年的里约热内卢城市旧城改造政策的结果,虽然在本质上截然不同——无论是开发目标、环境及管理背景方面,里约港区复兴计划与古根海姆博物馆项目的组合都体现了国际盛行的一个趋势,即将大都市的废弃港区改造为商业、旅游、休闲娱乐区,像波士顿、巴塞罗那、伦敦、布宜诺斯艾利斯等大都市一样(图 7)。

(七)里约热内卢的港口

里约热内卢的港口是全国最重要的港口之一。它于 1910 年落成并受控于法国投资,1911~1922 年间由里约热内卢港口公司(la Compagnie du Port de Rio de Janeiro)代为管理。1923 年,它交还给国家当局并由巴西港口管理公司管理。1936 年,里约热内卢海港管理公司成立。1973 年,瓜纳巴拉船坞公司成立,并在 1975 年改名里约热内卢船坞公司。

从1993年起，港区的经营活动逐渐通过特许权转让的方式转移给私营公司，这些私营公司像欧洲大港的运营模式一样，提供终转港服务。这些变化使港区及城市金融中心附近的仓库腾出了更多的空间，港口当局拟在此建造购物中心、商务中心和文化中心，并改造港口操作区以使港口具有更现代化的船坞，并可以提高港口和其他区域的交通连接。

（八）里约热内卢港口的形态以及运作特征

里约热内卢港口位于瓜纳巴拉海湾的西岸，直接有高速公路通往圣保罗、贝洛哈里桑塔和圣萨尔瓦多的联邦高速公路，也有铁路连接里约热内卢州南部中心地区（Paraiba Valley），并可以从那里前往圣保罗和帕拉那州。同时，铁路向西北部地区可以通往Espirito Santo和吉赫金矿。港口堤坝宽度为1.5km，最小水深为12m，两边分别以Pão de Açúcar消防站和位于瓜纳巴拉湾入口处的Santa Cruz要塞为界。入口处的运河长18.5km，最小宽度和水深分别为150m和17m。

1. 设备设施

港口有总长度6740m的连续泊位和392m的防波堤，其分布如下：

Mauá码头长880m，拥有5个泊位，其深度从7~10m，泊位面积为38512m²。

Gamboa码头从Mauá码头起始并且延伸至Mangue运河，其长度为3150m。它有20个泊位，其深度从7~10m。它还有18个仓库，1个容量为15200t的冷库和60000m²的露天堆场。

Sao Cristovao码头长1525m，拥有6个深度自6~8.5m的泊位，两个总面积为12100m²的仓库，以及23000m²的露天场地。

Caju码头是一个可供车辆直接驶入的码头，其泊位长度为1001m，深度6~12m，3个仓库的总面积达到21000m²，并且露天堆场面积达69200m²。

集装箱码头装卸区总面积达到137000m²，包含公路和铁路出入口。装卸区长度为784m，有4个装卸点，另有280m长的码头，泊位面积达到324000m²。

另外，港口外还有10个仓库，总面积达到65367m²，8个带覆盖物的场地总面积为11027m²，总容量为1.31万t。码头上主要的货物为纸张、小麦、轿车和集装箱，码头外的货物主要有石油及其副产品。

（九）码头区复兴计划

1. 目标区域

码头区的全称是里约热内卢第一行政管理区。这个区域从Mauá广场（曾经是客运码头的旧址）开始，沿着瓜纳巴拉海湾的滨水岸线，经过港口的装卸区，直到Cajú居住区附近Niteroi桥的起点。复兴规划所包含的地区包括Rodrigues Alves大街及其他靠近市中心的地区，1~18号仓库也在其中。规划的第一阶段将改造1~6号仓库。

研究的区域包括码头、始建于20世纪初的大片城市区和早期海岸线旁的小山，其中比较著名的社区有Saude、Gamboa和Santo Cristo，这些地区的首字母组合起来就是SAGAS，也就是里约热内卢市中心区域古建筑及环境保护计划，这个计划始于1980年。

相对于其他港口区域的复兴工程，有一点值得明确的是整个里约热内卢港口完全是城市化地区。这个区域由非常独特的部分组成，包括坐落在装卸区内宽阔笔直大街上的大型仓库，还有那些位于古老海岸线和小山坡狭窄而弯曲的小路旁，具有葡萄牙城市风格的窄而深的低收入者住宅。在这些区域的中心，原先Gamboa的铁路站场已经变成一片巨大的空地，在过去的几年中，曾经有一些计划打算将这个区域改造为居住区或嘉年华会。另一个巨大的铁路废弃站场坐落于这个区域西部边缘的Santo Cristo山脚下。

这些平地上大部分的建筑物都是联合教会的财产和联邦财产，码头边上的高地是废弃的仓库，附近还有低收入者居住的社区。这些低收入者主要是码头工人，居住环境虽老，但不属于贫民区。

另外，里约热内卢的港区同世界上其他区域一样，从20世纪60年代开始就受到高架公路及其相关设施建设的影响。这些建设完全不考虑当地的城市和环境现状。邻近港区的São Cristóvão 地区拥有重要的历史文化建筑，现在却像一个被高架路包围并肆意横穿的小岛。港区复兴规划必须面对的一大挑战就是靠近甘伯（Gamboa）码头仓库的高架道路穿越 Rodrigues Alves 大街而将海洋与城市割裂开来。

按照规划编制人员的说法，里约港港区的复兴规划目前尚处于最初的准备阶段，规划提出的原则与行动计划集中于"既要阻止地区下滑的趋势，又能使土地利用适应该地区新的活动要求……不是以他们为中心，而是尽量适应公众行为方式的……关注土地使用规范化的教育，对住房、道路交通，对公共空间社会与经济品质的再提升……总而言之，规划要力图成为提升与刺激城市变革的战略动力。"

指导方针与具体行动的目标有：

(1) 吸引新的投资；
(2) 体现文化遗产的价值；
(3) 刺激新的开发利用；
(4) 消除邻里间的隔绝；
(5) 恢复环境质量；
(6) 重新整合瓜纳巴拉海湾区域；
(7) 强化居住功能；
(8) 在不同的层面上促进当地经济的发展；
(9) 为了实现这些目标，需要采取以下行动；
(10) 制定周详的规划；
(11) 建立一支项目管理团队；
(12) 寻找刺激当地经济发展的出路；
(13) 执行城市规范与规划；
(14) 重新设计道路及交通运输系统。

参与合作的社会机构有：

(1) 经济与社会发展银行（BNDES）和联邦政府援助银行（CEF）；与这两家银行的合同已经签署。经济与社会发展银行将参与轨道交通建设工程的开发研究；
(2) 联邦政府和州政府；
(3) 当地及国内外投资商；
(4) 当地的协会；
(5) 文化及环境领域的非政府组织（NGOs）；
(6) 国际发展机构。

（十）文化遗产保护与城市规划规定的调整

现行的城市规划仍然遵循着传统的功能区划原则，为保证港区使用及各项活动的进行，严格限制居住使用。里约港区规划的要点包括：

(1) 优先考虑混合使用的功能区；
(2) 鼓励与完善居住区的发展；
(3) 沿着甘伯码头（Gamboa Quay）形成一个满足多样化活动需要的"功能核心"，集合商业、服务、文化和休闲等多种功能；
(4) 扩大建筑文化遗产保护的清单（2000年），以涵盖那些建于20世纪初具有历史和建筑保护价值的建筑。

（十一）道路、交通和运输系统

规划目的在于使内部与外部之间、港区与城市中心和其他地区之间的联系更加清晰，具体项目包括：

(1) 一个综合、连续、分等级的地面道路规划：①建造一条不受高架路影响的新的内部干道，为此将重新使用一部分废弃的铁路系统；②街区规模以及由巨大的仓库和铁路堆场所定义的城市尺度将重新定义；③沿着山脚的路形成了曾经的海岸线，因此将被保留并且完善。

(2) 一个新的轨道交通系统（VTL），它将连接所有市区交通系统的终点站和各历史景点，包括公交总站、勒奥珀蒂那车站（Leopoldina）、Estácio 地铁站、中央车站、Cruz Vermelha 广场、拉帕凯旋门（Arcos de Lapa）、Cinelândia、圣杜蒙特机场（国内航班）、第十五广场和 Mauá 广场。

（3）一条沿着港区的自行车道把南部海滨与城市中心的自行车道系统连在一起，也将把港区和未来的古根海姆博物馆、第十五广场和现代艺术博物馆连接起来。

（十二）住房

本计划将增加港口区的人口数量，从现在的 2.2 万人增加到 4.2 万，且其收入达到国家最低工资（大约 550 美元/月）的 10 倍。为了达到这一目标，规划将：

（1）激活并赋予城市中大块空地以新的功能；

（2）充分利用地区内的小型闲置地块，特别是地势较高的地块；

（3）修复那些正在衰败的建筑；

（4）改造该区域内现存的 3 个贫民区。

（十三）公共空间

基于长期积累的历史文化保护以及商业街复兴工程的经验，规划将：

（1）重整街道和广场体系；

（2）恢复历史性环境；

图 8　Conceicao 山历史街区的一角
（来源：P. Jorgenssen Jr.）

（3）拆除所有影响从海湾到山顶视线的围墙和其他构筑物，建设小型公共广场和公共走廊。

（十四）增强当地的经济实力

具体做法如下：

（1）加快建筑执照审批程序；

（2）建立经济开发区，以暂时减免税来吸引投资；

（3）对本地企业提供信用及小额信用额度支持。

（十五）主要投资项目

1. Mauá 广场——6 号仓库

Mauá 广场重建项目的投标工作已经开始，其内容是在 Rodrigues Alves 大街与 6 号仓库之间建设新的城市文化中心。为了达到这个目标，将建设拥有上千个车位的地下停车库和一条自行车道，现有的联邦警察大楼将改造成为购物中心。周边高架路下方的柱子将做景观化处理，以融入未来的新环境。古根海姆基金会已经委任法国建筑师让·努韦尔在 Mauá 码头设计新的古根海姆博物馆，并指导 Mauá 广场重建项目建设公司的工作。除了适应市中心不断增长的停车位需求外，该停车场也将满足未来新博物馆的停车需求。工程预计在 2004 年完成，其项目获得的土地使用权是 35 年。

2. Morro da Conceição 地区

从 Morro da Conceição 山脚下延伸出去，一个方向是港口区，另一个方向则是金融中心。有关这一地区历史与环境保护的详细提案在 5 年前就制定完成，现在已在法国顾问的协助下由市政府组织的专家组实施（图 8）。

3. Barão de Tefé 大道及邻近区域

这个街区云集了大量的高科技公司，国家科技研究所也在此处。这项计划将建造地下车库，并对地面部分进行更新，主要是公共照明和绿化方面。

4. Saúde 山和教堂

Saúde 山是港口区的一个重要历史景点，有始建于 1742 年的 Saúde 教堂，已经落实政府拨款予以修复。由于这座山是靠近海岸线的最高点，它被选为未来客运港的重要对景之一，整个工程将耗资 30 亿雷亚尔（约合 13.3 亿美元），历时 12～15 年。

5. 古根海姆博物馆

尽管在市政当局与建筑师协会、艺术家及其他博物馆之间存在争议，位于港区的古根海姆博物馆工程已经进入建设的最后阶段。该工程由法国建筑师让·努韦尔设计，正式开工选在 2003 年 1 月，开工同时还在里约港和纽约市举行了两个博览会。鉴于它继承了原有圣彼得堡博物馆的馆藏，新博物馆将被冠名为古根海姆里约博物馆。

市政府耗资 200 万美元进行了项目的可行性研究，并希望在 2003 年通过招标投标的方式来建设。在三年的建设期里，公共投资将达到 2 亿美元，其中包括 2000 万美元用于古根海姆这个商标的使用费，合同还规定在博物馆未来的固定收入中将有一定比例用于支付给古根海姆商标所有人作为版税（图 9）。

建设区域沿着 Praça Mauá 的 1 号码头展开，占地面积 2.1 万 m^2，呈一个长 400m 宽 80m 的矩形。由于一大部分的结构隐藏了起来，这个方案看起来就像码头边船只的停泊点。用来展示历史艺术、巴西当代艺术、多媒体艺术的不同空间是半地下结构。它的顶部是圆锥形，覆有玻璃天窗来满足自然采光。穿过一个迷你的热带森林，游客将进入到海平面下的部分。在码头的另一端，一个巨大而外表腐蚀过的圆柱形塔作为结束，这是永久展览、实验艺术的展区以及设备区。在它的顶端 42m 的高处，将有一个观景平台和全景餐厅（图 10）。

据初步估计，每年将有 100 万游客前来参观，直接收入达 400 万美元，同时，每年将给城市其他地方带来 7500 万美元的间接收入以及 600 万美元的公共税收。项目将增加 5000 个就业岗位，港区内及周边地区房地产的价格将上涨 50%。据预测，该项目还将带动周边的私人投资，如宾馆、会议中心、购物中心、音乐厅和其他配套设施。

其他社会组织如何从中获益目前仍在探讨中，预计每年将要投资 1200 万美元用于增加馆藏艺术品，首先将考虑巴西的艺术品，其次是拉美的艺术品，此外还将从圣彼得堡及古根海姆博物馆中转移部分藏品。博物馆将由名叫里约艺术基金会的顾问组织与古根海姆基金会共同管理。

图 9　古根海姆里约博物馆
（设计：Jean Nouvel）

图 10　城市海岸线

七、结论

对里约港口的规划倡议包括了城市、省及联邦政府以及私人机构，其共同的努力将带领港口区域迈向复兴与具有活力的新时期。同时，对于 2007 年泛美运动会的后评估表明，全球性的娱乐、旅游及商务中心活动及城市发展间的密切联系，因此，城市可以学会有效地利用这些活动给城市带来的形态、社会及社区发展方面的提升。同时，一个残酷的教训来自古根海姆博物馆的案例，确认及接受私人机构对城市大规模项目的推动作用，并不意味着公众舆论监督和公共控制就能因为所谓的商业秘密或国家原因而置之度外。

城市大规模项目的建设因为涉及一系列跨学科的知识与技能，为快速推进城市规划工作的展开及规划文化的建设提供了难得的机会，但是，更为重要的是有效而民主的城市管理实践。

参考书目

[1] Abreu Mauricio. Evolucao Urbana do Rio de Janeiro, 3a. ed. Rio de Janeiro: Instituto Pereira Passos [Z], 1997.

[2] Centro de Informações e Dados do Rio de Janeiro – CIDE, Anuario Estatistico [Z], 2001.

[3] Companhia Docas do Rio de Janeiro[Z/OL]. www.portosrio.gov.br.

[4] Diário Oficial do Município do Rio de Janeiro [Z].

[5] Instituto Pereira Passos. Armazom de Dados [Z/OL]. www.armazemdedados.rio.rj.gov.br.

[6] Instituto Pereira Passos[Z/OL], www.rio.rj.gov.br/ipp.

[7] Jorgensen Pedro, Rabha Nina. Rio de Janeiro – Recovery and Revitalization, the City and Its Port [J]// Carmona Marisa, ed. Globalization, Urban Form and Governance no.10, Delft: Delft University of Technology.

[8] Prefeitura da Cidade do Rio de Janeiro. Porto do Rio-Plano de Recuperacao e Revitalizacao da Regioo Portuaria[Z]. Instituto Pereira Passos (IPP), October 2001.

RIO DE JANEIRO

Pedro Jorgensen

City History

The city of San Sebastian de Rio de Janeiro was founded in 1567 as a fortress. In the XVIII, the gold of Mina Gerais, displace the colonial activity axis to the southwest and in 1763, it became capital of the Viceroyalty. Slaves and a small merchant, military and administrative elite formed the population. In 1808, the transference of the court (around 15000 persons) threatened by the Napoleon's incursions, transform the colonial capital in seat of the Portuguese government and economical centre of the big coffee region from Southwest Brazil. The opening of ports breaks the barrier of colonial monopoly and the city suffers important transformations. In 1822, the independence makes Rio, the capital of Brazilian Empire, the population increase, it arrive foreigner capitals and mechanic transport start to trace the modern city.

Since 1850 the public services where given in concession to foreign companies. The railway development stimulate the formation of suburbia that later will surpass the urban borders and will configure the metropolitan space. In 1868, tramways of animal traction give the transport service toward the new urbanizations in the coastal border and in the mountains to those that could afford it. In 1900 Rio already had 500000 inhabitants and was consolidated as capital of an agrarian republic (1889) ruled by a coffee bourgeoisie. The need of modernization implied actions in two fields, the first was the urban sanitation and the second was the *haussmanian* urban transformations. As a result of these actions, a contingent of poor population increased by the slaves emancipation (1888), is removed from the centre to build the central avenue. A new port is inaugurated in 1910.

During the years 1920 the hill *Morro do Castelo* is removed combining the *Beaux-Arts* and *City Beautifull* theories with elements of urban engineering. Those parameters are also maintained during the Vargas age (1930-1945), that create the basis of the modern Brazil with the state regulation of the relations labour-capital, born from the industrialization and the emerging of the industrial worker and the middle urban class. In 1940 is open the Brazil avenue, big way of access to the city, and the *Presidente Vargas*, monumental axis of 100m. wide framed by high building on "galleries".

From 1940 to 1960 the city is transformed into a metropolis. The State supported industrialisation and the rural exodus originate a vast periphery without public services. The city population goes from 1.8 to 3.3 millions in the city and from 2.2 to 4.9 millions in the urban agglomeration. The number of favelas [1], the

---

[1] Neighborhoods of spontaneous formation in which the poorest urban population is settled. Shantytowns.

traffic congestion and the shortages of drinkable water are increased. The car industry becomes the economic motor and the symbol of modernization. The road development and the systematic clearance of favelas, become the core of the urban management. The tramways are replaced by buses and cars; long tunnels and large residential areas in the periphery, are built.

The transference of the capital to Brasilia did not affect Rio de Janeiro. Its importance as a metropolis is due to the port, the remaining entities from the federal administration and its industrial and services park. In 1975, Rio became capital of the new State of Rio de Janeiro. The Urban Basic Plan (PUB-RIO), the integrated plan of transport and the Lucio Costa Plan for the Baixada de Jacarepaguá, contributed to extend the city through their peripheral shopping's malls and motorization.

The recession, the debt crisis and the democratic mobilization for the Constitutional reform of 1998 create the conditions for the municipal autonomy. Experiences of favelas upgrading and of urban conservation in the central area arise. The Ten Years Director Plan (1992) was proposed to promote a democratic urban management based on equity and some legal mechanisms begin to be applied in urban regulations and in the upgrading of favelas in the 1990's, but the created ground will not be regularized.

The democratic movement of the 1990's coincide with a low economic growth, the increase of inequality and a general informality of economy. The adopted urban strategy combine public focalised targeted investments of big impact (Favela-bairro, Rio-Cidade) with a new generation of urban services concession and the strategic planning oriented toward the "competitive integration" of the city to the world market, through development of tourism and the celebration of large events (Teleport, Waterfront, 2004 Olympiads). The millennium starts with the port revitalization proposals, the Pan-American Olympic of 2007 and the Guggenheim Museum project.

Globalisation

The city has already a history of competitive integration to the world market of good and services of planning and urban projects. A new focus of the urban project toward the global market starts in the 1980's with the idea of transforming the port area in a teleport. An "urban management culture" arouse since the seminar "Urban Strategies Rio-Barcelona" celebrated in 1993. Since then, the city has been developing plans, studies and projects oriented toward the exploiting of competitive advantage. That processes will be summarized in the Program Urban Strategies and Urban Projects, being the Revitalization of the Port Area / Guggenheim Museum, the most important project.

It is a fact that the economic globalisation created by the central economic blocks, the asymmetric opening of markets and the advances in telecommunications affected the urban management during the 1980's. The synthesis of this changing perspective is the emergence of the "competitive city", which is able to attract to his territory investments of multinational enterprises (industrial and financial), mega-events and tourism. Metropolitan centres became the fundamental generators of real state projects and related businesses, enough profitable to pay for they infrastructure and urbanization.

The main task of the urban management – oriented toward and by the globalisation – is constituted by the reconversion projects of the obsolescent industrial sites, port, railways an other infrastructures into new business centres, industrial parks, historic-cultural tourism and gentrified housing areas, generally located in central areas of large metropolises and their surroundings, but that could be located as well in vital transport and circulation nodes.

The large urban projects could become a field of application and development of effective and transparent techniques of public administration. On the other hand, one of the impacts of economical globalisation is the verticalisation and lost of managerial capacity of the public administration. The processes of negotiation and decision making of large urban projects are done in circle every time smaller and closed to the public scrutiny.

Urban Strategies

In 2003 Rio de Janeiro strategic planning process will complete 10 years. The director plan of the city is a legal code designed to promote an urban management of redistribution that express the thinking of a democratic movement that latter, in 1998, generates the mobilization for the new constitution. As a consequence, although they are oriented to complementary aspect of the urban management – normative and executive – the Director Plan and the Strategic Plan contain conflictive visions of public policies regarding the objectives, the main agents, and the methods.

The urban strategies effectively applied during the period can be described in four points:

1. Urban development oriented toward improve city competitiveness: Teleport construction, concession of *Linha Amarela* (yellow Line) Olympic candidature, Strategic plan, International competition to the concession of urban furniture, emblematic proposal for the inner-city revitalization, port revitalization, Guggenheim Museum, Pan-American Games 2007.

2. Institutional and normative flexibilisation: Creation of a Plan (Strategic) and a politic representation (Urban Council) alternative to the State Institutions, constituted by social agents with capacity of investment and opinion formation, reiterative use of "Interlinked Operations".[①]

3. Emblematic Public investments to promote positive socio-economical and cultural effects: programs *Favela-Bairro* and *Rio-Cidade*, oriented toward the strengthening of household economy and small enterprise at local level in addition of meeting the strategic objective to stimulate the city self-image. The impact of those programs is now well known in the city, in the country and abroad.

4. City marketing through "virtual projects": Some projects are not presented as public action designed to attend basic needs, but as an idea to attract investment. The architectural image performs the double role of promotional object and of feasibility test, as is the case of the publication of a luxurious photomontage with the downtown area projects. It is clear meanwhile that a set of "feasible buildings" does not make it necessarily an intelligible Plan.

Urban Projects

The Teleport

The teleport is the business district equipped with last generation technology of telecommunication. With

---

① This instrument allows – through a recommendation of an experts commission of urbanists, approved by the Mayor and sanctioned by the City Council – The purchase, by a particular entrepreneur, of rights of use and occupation of ground, additional to the established in the normative. The income of the I. O. are destined, via Urban Development Found, to the improvement of *favelas*.

a lot area of 250.000m², 62.000m² built and 450.000m² of total space use, this area contain the main municipal central administration and a Metro station. It is located close to the financial centre and to the airports: the international one and the domestic one, and to the port. The investment has already reach US$ 130 millions and it is expected US$ 260 million additionally in the next three years.

Praça XV de Novembro Restructuring

Located in the waterfront of Guanabara Bay, this square is the epicentre of an historic core of 1600 restored buildings. The restructuring consisted on the demolition of an elevated pedestrian viaduct, the construction of an underground way and the redesign of the public space. A second phase of the project has been promoted in 1998 with the transformation of the square waterfront (1km) in an area of public and cultural activities.

Linha Amarela (Yellow Line)

It is an express way of 25km that connects the *Barra da Tijuca* (area of Real State expansion) with the international airport and the metropolitan highways. This is the first urban highway with toll in Brazil, and was executed mainly with private investment, within the concession regime. The main weak points of the project are: the exclusive destination to private transport and the non realization of re-urbanization projects in the adjacent expropriated areas which have been immediately occupied by informal settlements.

Favela-Bairro Program

The objective of this program is to integrate the *favelas* to the adjacent neighborhood with complementary investments to the inhabitant's self effort in the construction of their dwelling through provision of sewerage, social equipments and income generation programmes. The programme has being paid with municipal resources, federal credits and IDB credits, in addition to non-refundable resources from the European Union. The program covers 104 communities.

Rio-Cidade Program

This program consist of the recovering of deteriorated public spaces due to the construction of the new road infrastructure developed during the 50's and 60's and the develop of exclusive spaces – Shopping Centres and gated communities – during the 80's and 90's. The strategy consists of the generation of multiplying effects by means of emblematic focalise actions.17 commercial streets as priority interventions have been chosen.

The Large Urban project: Plan of recovery and revitalisation of the Port Zone

The recovery and revitalisation plan of the port area and the construction of the Guggenheim Museum are semi-independent initiatives but complementary within the municipal realm. The location proposed for the future museum – the pier of *Praça Mauá* – strategically located in the crossing of the lines formed by the port quay and the Rio Branco Avenue (the financial centre of the city), is considered the most important valorisation zone in the port area. The Municipal Institute of Urbanism Pereira Passos IPP, is in charge of the development plan for the port area.

The port of Rio de Janeiro

That project was born from people's reaction transform the port area into a "Teleport".

In the heart of that territory, the old Railway yard of Gamboa constitute a big empty space that during the last years has been object of several usage proposals. A second obsolete railway yard of large dimensions is located on the west border of the area, at the base of the Santo Cristo hill. The Port Area Recuperation and Revitalisation Plan – Porto do Rio – is in its preliminary stage of elaboration as a Director plan.

The Guggenheim Museum

Despite some polemics between municipal government circuits, the Institute of Architects, artists and responsible for the museums, the Guggenheim project in the port area (Praca Mauá) is in its final stage of elaboration for which the French architect Jean Nouvel has been appointed. It is expected to be officially presented in two simultaneous exhibitions in Rio's Port and in New York in January 2003. The museum of Rio will be named *Guggenheim/Hermitage Rio*, inheriting in this way the association of origin with Saint Petersburg's Museum. The feasibility study has cost US$ 2 millions to the municipality. The estimated public investment is US$ 200 millions for the three years of building works, including the US$ 20 millions for the Guggenheim trademark license. In addition, the contract sets forth payment of royalties on a percentage base of the future museum's regular incomes.

罗萨里奥（阿根廷）

图1 罗萨里奥在阿根廷的区位图

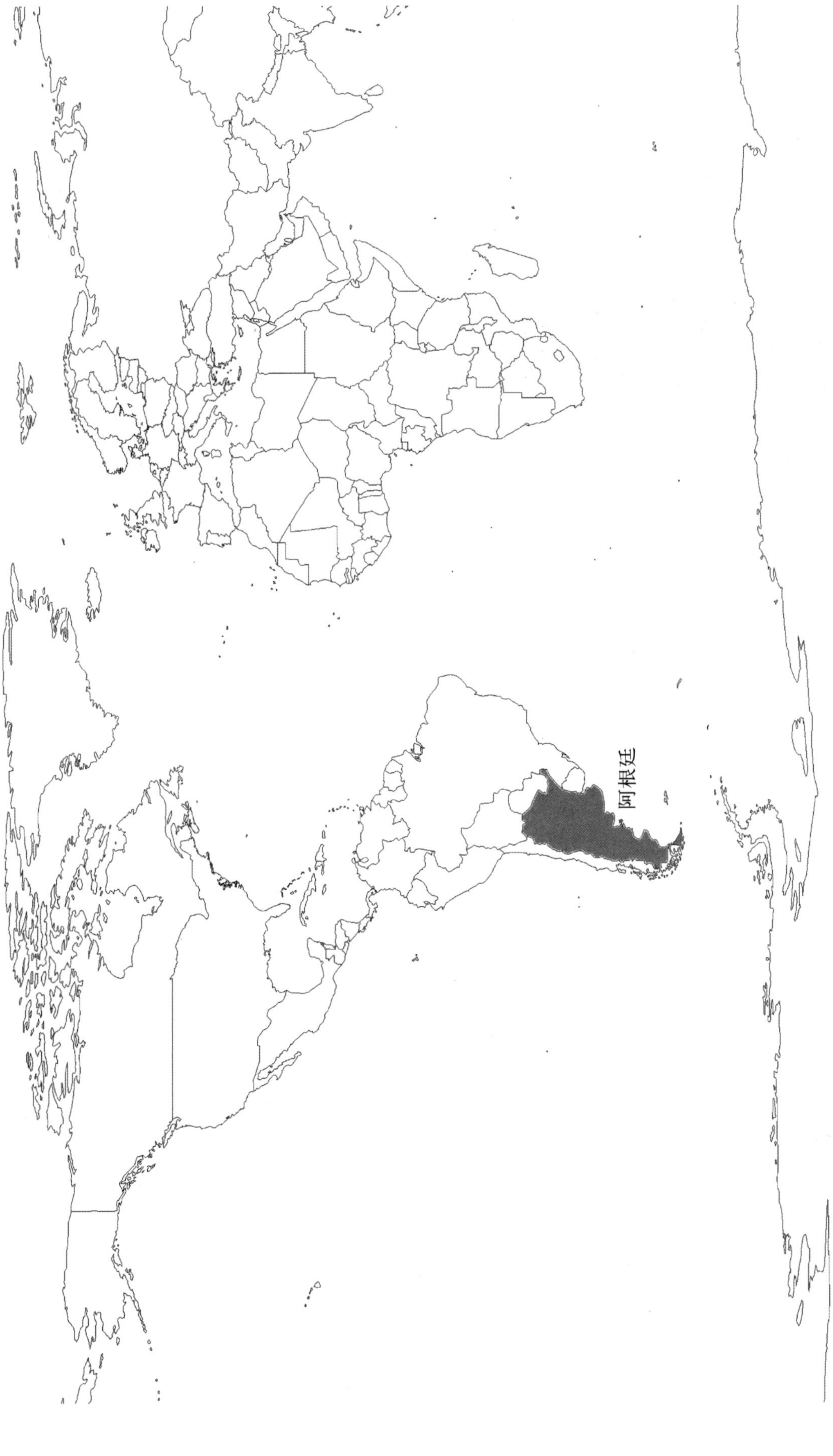

图 2　阿根廷的全球位置图

罗萨里奥

罗伯托·蒙特沃德 奥斯卡·布拉格斯

一、城市数据

（一）罗萨里奥市市域范围

总人口（2001年）：1162029人。

两次普查间人口增长（1991～2001年）：3.6%。

面积：48800hm^2。

罗萨里奥市域城镇：Puerto San Martín、San Lorenzo、Fray Luis Beltrán、Capitán Bermúdez、Granadero Baigorria、Villa Gobernador Gálvez、Pérez、Soldini、Funesy Roldán。

劳动人口比例（2001年）：43.8%。

就业率（2001年）：33.8%。

失业率（2001年）：22.8%。

未充分就业率（2001年）：17.2%。

从事经济活动的人口（2001年）：565642人。

职业分类（依照不同的活动）（2001年）：

工业：74007人。

建筑业：32137人。

贸易：96905人。

教育：37556人。

家政职员：38952人。

人口结构（2001年）：

雇主：22636人。

自由职业：119754人。

雇员：285498人。

失业人员：8582人。

（二）罗萨里奥市

人口（2001年）：907884人。

1991～2001年的人口增长：-0.1%。

面积：17869hm^2。

住房（1998年）：

自有：64.5%。

租用：14.9%。

空闲：7.4%。

其他：13.2%。

（三）住房

有给水系统住房（1991年）：91.4%。

有排水系统住房（1991年）：55.4%。

有供电系统住房（1998年）：96.0%。

有电话的住房（1993年）：45.5%。

生活用水量（1991年）：每人每天171.73L。

新生儿死亡率（1993年）：2.33%。

犯罪率（1993年）：

谋杀率：0.06。

偷盗率：3.32。

（四）交通与道路

机动车数量（1999年）：345853辆。

公共汽车（2001年）：617+20辆无轨电车。

公交公司（2001年）：9+1家（无轨电车）。

每月客流量（2001年）：8053916人次/月。

每月平均每辆公共汽车乘客数（2001年）：13053人次。

出租汽车（2001年）：3127辆。

街道铺装率（2000年）：85%。

绿地/开放空间比例（2000年）：5%。

人均绿地/开放空间面积（2000年）：9m²/人。

大学毕业生（1999年）：占总人口的7.08%。

电影院（2000年）：31间。

港口货运量（1998年）：2384724t。

航空旅客数（1999年）：353003名乘客/年。

（来源：常住居民民意调查、省统计普查研究所、罗萨里奥市议会、罗萨里奥战略规划）

二、城市概况

作为阿根廷唯一的非省会主要城市，罗萨里奥坐落于圣达菲省的南部，人口约为该省省会的2.5倍，全省30%的人口集中于此（图3、图4）。

罗萨里奥于18世纪设市，但建市却直到19世纪末才开始。贸易的繁荣成就了罗萨里奥高水准的城市基础设施建设服务，虽历经前几年的经济萧条，这种服务仍得以延续。

罗萨里奥的城市化模式是低密度的，其原因是市内大多数房屋都是独幢房。人群最为密集的建筑群集中在市中心（集中了大约15%的居民）。与其他的拉丁美洲城市不同，罗萨里奥的市中心集贸易、金融中心与住宅区于一身。尽管这些特点使这片区域更具活力，但在过去的几年中，区内的商业已现出疲态，开始淡出人们的视线（图5）。

市内居民主要是西班牙和意大利移民的后裔，以及一些黎巴嫩人和来自东欧的犹太人后裔。过去几年里，从阿根廷北部搬迁而来的人也使得城市种族更具多样性，Tobas就是其中一例。由于这里有完善的公

图3　市域图

图4　市旁的Cordoba山

图5　市中心的街道格局

众医疗体系及社会政策优待，许多阿根廷北部贫困地区的人都相继搬来这里。

小公司推动了整体产业的发展，其中最重要的是纺织、制药、塑料、电器、家具和食品业。

三、城市历史

长久以来，罗萨里奥一直是阿根廷最重要的城市之一。位于 Paraná 河畔，其得天独厚的地理位置与其经济发展有着密切联系。鼓励农产品出口并鼓励移民和殖民的政策，使罗萨里奥成为谷类食品生产区的中心，并占据全球市场的一席之地。而在 19 世纪中期前，罗萨里奥只是个地处边境、人烟稀少的小村庄。后来，铁路和港口的建设以及欧洲移民很快地改变了这个小村庄居民的生活，也提高了这个村庄的影响力。

因此，城市成为北部、中部及东部地区产品的输出港，50 多年来城市的人口增长了数倍。因为城市开始发展，就有了自然环境和生产活动之间、开放空间和规划控制区之间的矛盾。由于汇集了所有港口和铁路终点站，河岸就成了矛盾汇集的焦点。历次规划都提到河流与河岸的问题，很多提议倡导迁移港口和铁路终点站，把河岸周边的区域作为公众使用的公园。

1873 年，罗萨里奥有了第一个扩建计划。这份计划旨在发展这个有着 3 万人口的城市。计划中，河岸被视为城市最重要的地点之一。于是扩建计划中就有了在峡谷里建设大道的提案，大道各端风景优美，是人们散步的理想场所。因此，一开始峡谷就被视为公众休闲场所。

数年之后，在 1910 年负责 d'Aménagément 计划的法国建筑师 A. Bouvard 重新考虑了这个想法。这次，他们计划通过峡谷上的林荫大道连接各公园，宛如一条绿色的丝带，分隔开城区和港口，却又在视觉上连接了城市与河流（图 6）。

在 20 世纪 20 年代，城市美化运动出现。在那时，罗萨里奥市人口超过 250 万人，城市建成区已经向北方、南方和西方延伸，其港口已经成为国内第二大港口。新的需求要求规模与品质的提高，这就需要罗萨里奥转变为一个现代化的城市，城市既要有高效率又要赏心悦目。在商人和地主的鼓吹下，这次城市美化运动的目标之一，就是把港口和铁路设施迁出市中心。Sociedad de Ingenieros（工程师协会）警告人们，这种做法（把港口和铁路系统迁出市中心）将会威胁到城市最重要的发展。但是那些提出城市美化计划的人坚持要这样做，为了达成目的，他们提出了新的工具，即城市总体规划。因此，1928 年，罗萨里奥聘请了三名本地工程师重新策划，新的计划于 1935 年提交公众讨论。

新计划依然纳入了美化城市运动的目标——将港口迁至建成区以南，留出市中心和北部的河岸用于建造公园、花园及河滩。这个计划雄心勃勃：把河岸建造成一个适合娱乐、休闲和运动的场所。从 1935 年直到现在共有四轮不同的规划，它们全都贯彻了为百姓造河岸的思想。

1953 年的规划以重新设计铁路系统为基础。经历战乱之后，国家的经济和政治情况已经改变，包括工业化、铁路公司国有化、国内居民大迁移。罗萨里奥也经历了这一系列变化，人口数接近百万，周边的村庄也被城市化，成为了城市连绵区的一部分。因此，规划中希望将铁路迁出河岸，只留下港口。

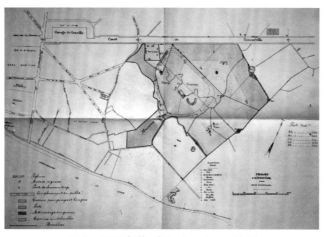

图 6　1910 年的 d'Aménagément 计划

1968年制定了新规划,同样希望将河岸还给公众。它的基本思想与1935年的计划相同：将港口从北方移动到南方，并且在中心河岸建造公园。

1991年制定的新规划中依然保留了旧的思想，河岸规划为三个部分：北部用于运动和娱乐活动；中心用于娱乐和文化活动；南部用于港口活动。但是，与早先的提议不同，港口仍留在中心区北部，而以谷类输出的港口转变为集装箱港口。考虑到城市建成区的发展和整合，这并不是一个恰当的提议，好在2000年的规划进行了调整。将河岸转变为公共空间的规划始于百年前，而今正逐步变为现实。

四、全球化，机遇与挑战

经济的全球化以及行政结构的改变给圣达菲省的土地组织带来巨大变革，位于其南部海岸的罗萨里奥处在变化更为明显的地域。今天的土地组织和十年前有许多差别，而且有证据表明变化会越来越大，将使罗萨里奥今后在本地区发挥全新的作用。

许多因素的共同作用造成了今天的局面，包括：公共政策的变化（享受国家大幅补贴的国有企业改制标志着国家福利的终结、公共服务私有化，以及闲置土地转交给镇议会）、经济上的变化（工业结构调整、社会服务发展、金融活动过度管制、失业增多）、区域组织上的变化（由科尔多瓦省、圣达菲省和恩特雷里奥斯省共同组建中央区，阿根廷加入南美国家共同体），以及社会动态学上的变化（国内移民、贫穷人数和社会排斥度激增）。这些变化的过程，体现了罗萨里奥及其所在地区正在改变的特征，尤其是基础设施的现代化、工业结构调整、城市化的新形式以及社会不平等的加剧。

随着象征解散国有企业的本国法律生效，又有了新的关于Paraná河滨区的计划。终结国家对于港口活动的垄断标志着Paraná沿岸各种港口网络的形成，固体货物、矿产、石油和化学产品输出港相继建成。

此外，不久的将来，罗萨里奥及其周边都市区的基础设施会有显著的改变，这些正在进行的项目将从根本上改变城市在本区域的地位，连接罗萨里奥和科尔多瓦的高速公路以及连接罗萨里奥和维多利亚的大桥建成后，罗萨里奥将成为连接巴西南部和智利而跨越两大洋的通道。

这个计划前景美好，与强烈冲击都市区工业的金融危机形成鲜明对照。这个阶段的工业结构变化常见的特征有就业岗位短缺，许多企业停业的同时也有很多新开张，制造业开始出现投资成本国际化。城市北部的中心在历史上工业化程度最高，也是受此过程影响最大的区域。在20世纪90年代中期，机械工业和小型化工业消失了，伴随的是20世纪80年代中期开始的农产品加工业的蓬勃发展（尤其是榨油的企业），其产品大多出口国外。

一方面，南部地区也经历了类似的变革，一些传统从事肉类加工的小型食品和化工企业相继倒闭，取而代之的是生产汽车零件的大型工业集团。

另一方面，在过去的几年中，在罗萨里奥城市近郊区域已经出现了土地利用的新方式。与传统的将土地划分为小块、开发用途主要是社会性住宅的做法不同，新的居住和商业活动的模式已经产生，并与旧的方式同时存在。

中等或高收入人群生活方式的改变及通信手段的发展，以及房地产业者的策略等因素引导了住房市场的新变化。这样，居民除了私人社区和西部的中心外，还有更多居住方式可供选择。此外，过去十年中出现的新的贸易方式不再需要企业位于市区，可达性的改善已促使城市近郊出现新的形态、功能和社会面貌。

在过去几年中，大量社会群体加速边缘化，居住在"非法居住地"的人口大量增加，社会底层人群被赶至城郊村庄。年复一年，这种城市土地占用方式不断增多，不仅表现在土地使用者与所有者之间的争端，整个环境也呈现不断恶化的趋势。

总而言之，所谓"全球化"引起的一系列变化，是在不确定因素（经济演变）和确定因素（社会排斥）以及挑战（机构如何适应变化的环境，尤其是市政当局）的对比氛围中不断发生的。

五、城市战略

直到20世纪的最后20年，罗萨里奥总能成功地与政府的主要策略保持一致。因此，当阿根廷决定作为农业和家畜类中心加入全球市场的时候，罗萨里奥变成了沿海区域主要的谷物输出港，在世界大战期间，港口则变成以进口为主，罗萨里奥成为国家重要的冶金中心，城市和周边建立了许多工业中心。

20世纪末影响整个国家的危机也影响了罗萨里奥，城市的传统形象开始模糊，如何转变以应对新的形势？如何利用在建的重大基础设施，使罗萨里奥充分利用它的地域优势？地方政府开始尝试解决这些问题，聚集了经济、社会及文化领域的专业人士重新制定发展战略。1996年，组成了罗萨里奥战略规划委员会，并于1998年提出了最终方案。

战略规划的目的，一方面要确定哪些是根本性问题，以及找出城市的有利条件。另一方面，要选择、推荐关乎城市新形象的重要工程和项目，它们将成为城市转变的象征、城市复兴的标志。

整体上说，战略规划欲充分利用城市的新地缘战略地位，以此来改善都市生活质量，增强民众的凝聚力，强化城市与河流的联系，并把河岸建成为激励变化的中心。在具体工程中，有城际连接枢纽建造计划，如罗萨里奥与科尔多瓦间的高速公路、罗萨里奥-维多利亚桥（均在建），以及新的城市出入口，有提高公众空间品质的计划（绿化的区域分布、河岸复原计划），有改善经济基础设施的计划（港口和飞机场重建），甚至还有帮助市政府管理机构现代化的建议（分权计划）。许多计划已开始运作，重新树立城市形象。整个河岸地区都是大片的公园，极大地改善了城市形象，只有技术中心、出口协会、区域开发机构这几栋建筑掩映在绿化中（图7）。

（一）地方战略管理中的城市问题

依照这些工作要求，为了未来的发展，城市已经开始规划蓝图，寻找新的发展项目和新的管理工具。因此，城市项目必须成为总体战略的一部分，以使这个战略在新的城市综合管理模型中体现出国家政策的现代化、新的社会融合政策，以及城市项目本身。

战略规划的可靠性关键在于其本质，因其必须包含不同人群各自的主张，而地方空间的重构则需要大家对城市发展的规划达成共识。

全球化的要求从整体上引导本地在城市管理中的三种行为：社会规划、城市项目和现代化规划，这三者互相影响，并对战略规划起到反馈作用。因此，每个计划的提出都会触及社会、经济和都市问题，项目的提出是基于这些角度。在许多案例里，项目成功地实现了目标，其他的项目也许失败了。

六、城市大规模开发项目

（一）现代化和分权计划

这一计划的目标是提高行政效率，使市民更接近城市议会及其决策，这表示各区将根据其各自的特点重新审定城市政策，鼓励居民参与决策。现在，城市被划分为六个区域，共有三个区级事务中心，这些中

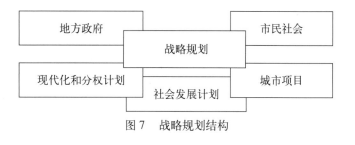

图7 战略规划结构

心组织会议，使市民参与、讨论、选择、审核通过那些改善生活质量的工程项目。

1. 非法聚居地改造的全面行动计划

这一计划的目标是通过包括新建住房、城市设施修复和社会整合来全面地解决非法聚居地的问题。为此，城市议会已经开展一个叫做"罗萨里奥栖息地"的特殊项目，而且已经得到泛美开发银行的融资支持。

2. 罗萨里奥—维多利亚大桥

这项工程连通罗萨里奥和维多利亚（位于Entre Ríos省），它越过Paraná河和Alto Delta群岛，总长度59.5km，包括长12.4km的桥和长47.1km的路堤。一方面，它将是美洲国家共同体中最重要的跨大洋通道，可以替代连接圣达菲与巴拉那以及位于布宜诺斯艾利斯的Zárate Brazo Largo隧道。另一方面，罗萨里奥将强化其作为最重要农业中心的地位，以及联系国际市场的主要枢纽。大桥2003年5月20日开放使用。

3. 河岸重建项目

整体而言，这一系列计划是城市与河流间地带的转变以及河滨建筑的一部分。由于它的自然条件和重要性，它们连同罗萨里奥—维多利亚大桥一起，成为城市在深受经济危机影响后振作及获得新生的标志。这些项目提供整个都市新的公共空间，以及新的文化和娱乐设施。

4. 罗萨里奥—科尔多瓦高速公路

项目的目标是改善科尔多瓦（位于阿根廷中部的一个城市）和罗萨里奥之间现有的公路，使之成为国家中心区域的主动脉。最初的一段（罗萨里奥到Roldán，长17.65km）于1999年10月竣工，整个工程则在2007年10月才全部完成。

5. 科技园区

这一计划的目标是为技术型公司的发展提供合适的环境。为此，在城市中选中的位置应该有优良的环境和基础设施条件，并与研究机构有直接的联系。技术中心已经聚集了许多计算机公司，正在测试科技园区发展的不同可能性。

6. 河岸重建规划

一旦Paraná河上的罗萨里奥—维多利亚大桥竣工，它将为城市向东部发展打开一扇门，把沿岸的区域和特有的风光与景致连成一体。

在河岸重建工程中展现的城市愿景，源于城市对河流的承诺：即河流和城市之间将建起一系列的绿色区域与公共建筑，以取代原先将城市与河流隔离开的旧铁路和港口设施。这个计划给不久以前还是难以接近、有着荒废工业区形象的地块带来了改变，赋予其新的角色，象征着城市向着期待中的模样变化。

在过去的十年中，有关改善城市和河流之间关系的计划和项目为数最多，其目标是为城市增加公共空间，而加强城市与河流之间联系的新方法包括增加生活性岸线（原来几乎没有）、增加使用功能（尤其是娱乐和文化用途）以及建立河滨的新外观。市民们多年来的夙愿将在21世纪头十年中实现——河岸不再用于铁路和港口设施，而转变为公民自由使用的休闲娱乐去处。

通过将港口迁到南面并拆除中心河岸的铁路和港口设施，可以腾出一块面积约90hm^{2}的区域，用于改建为设施用地和娱乐空间（图8）。

恢复计划基于三个原则，并确定了改变河岸各个部分（北岸、南岸和中心）的原则。

图8 Paraná河岸的改造成果

第一个原则是关于项目的区域规模，一方面要考虑到它的功能多样性，另一方面还要考虑其服务的都市及区域范围。事实上，在河岸出现的很多情况可以归结为这两个方面。

在过去十年里，旅游和娱乐业的发展迅速，尤其是分布着很多岛屿的北部河岸，水上运动和岛上娱乐业增长显著，其行政区属维多利亚镇政府管辖。

设在中部河滨的市中心发展计划，自从选择商业、文化和娱乐业后，新的项目不断启动（图9）。

一些计划的范围横跨几个地域，比如建在南岸的大型公园以及规划在北岸桥头的另一个公园。所有这些项目都需要不同的城镇间达成共识，以推动岛上旅游事业的发展，同时也保护这些地方的环境。

同时在罗萨里奥北部和南部相邻城镇，进一步促进滨河居住区发展已经成为一个趋势，因此需要建设罗萨里奥与相邻城镇间的道路，并贯通这些区域间的河滨步道。

第二个原则是这个河岸属于公众共有，不仅有新完成的绿色区域以及对所有公众开放的设施，还包括河岸中心区域和南部区域以及大学的校园。公园面积的显著增加是城镇议会的决策，意在更好地实现城市民主化的生活。

最后，第三项原则涉及对河岸各部分区域的评价，尤其是南部的峡谷区域，它正对着罗萨里奥港。在这方面，重建的规划一方面考虑怎样利用极致的美景来建设住宅，另一方面，需要继续营造绿化空间网络，通过绿化带来缓和港区活动对周边城市环境的冲击。

（二）北岸项目

1. 位于罗萨里奥—维多利亚大桥桥头的地区公园

罗萨里奥将和 Granadero Baigorria 市议会共同执行这一计划，因为可以利用的最大一块用地在 Granadero Baigorria 管辖中。项目一方面将大桥与公

图9 PASEO-RIBERE-DO 项目

园间建立良好的联系，另一方面则需要把大桥与附近的居住区适当分离。

2. 北岸环境组织的特别规划

规划目标是保护环境的风景特征，明确娱乐和餐厅的用地，并制定适当的规则保护现有的各种不同景点。

（三）中央河岸项目

1. Scalabrini Ortiz 公园

它位于罗萨里奥第一条铁路线的修理区旧址。由于其地址和规模，它曾被视为市中心区和北部相邻地区之间的分隔带。现在，由于废弃的铁路用地归还城市，重建的程序顺利开始。具体功能包括铁路建筑（阿根廷中心铁路公司办公总部和一个技术学院），一个商业、文化和居住区，一个公园，以及一片铁路使用区域。目标是创造一个新的城市中心，并成为进入城市中心前的北大门。当道路工程完成后，商场的建设资金却被暂时冻结。

2. 北部港口和 Scalabrini Ortiz 居住区

这个地方有很高的景观价值，因为它的位置特别适合欣赏岛屿，以及中部与北部的河岸。港口设施一旦撤走，就能用于建设 Scalabrini Ortiz 居住区和新的活动。按照规划，它将建成一个大型的城市

中心，项目中包括为城市提供更多的开敞空间，为城市的第三产业、文化、居住和旅游娱乐活动维持良好的空间。

3. 社区公园

作为城市建筑和都市遗产的一部分，很多过去的老仓库也坐落于此。近来，它已经变成一个公园，并将向公众开放。它的规划将通过专门组织的设计竞赛优胜者来制定，在其边缘将开设一些餐厅。

4. 西班牙公园（向北延伸区）

这一区域的高处有一系列弃用的铁路设施，而低处的河岸有一些属于钓鱼俱乐部的小型建筑物。规划初稿中打算保留现有的钓鱼俱乐部，以适应产业升级过程中文化活动的发展需要。此外，它还包括在中央区域建设的行政总部办公室（它们用于城市议会分散和现代化项目），以及对那些原属于旧铁路管辖的建筑物实施重建（图10）。

5. 旗帜陵园

它是中央河岸上最大的部分，也是城市最重要的形象区域，与中心区域有着长远的联系。它很大的一部分用来作为娱乐用途，其他的区域则设置航海设施和餐馆，尽管这些地方已经很久没用于港口活动。一旦把土地的所有权赋予城市议会，这一区域将会有长远的改变。一些最新的工程包括当代表现艺术中心以及河站（紧邻渡河乘客站的餐馆和商场）。因为最后的规划还没有完成，时下正在发生的变化都是根据1996年城市议会的决策要求一步步实施的。

（四）南岸项目

1. 罗萨里奥大学中心

罗萨里奥大学中心地处紧邻Urquiza公园的斜坡上，正对着河流，一条大街把它与港口区隔开。大学中心内有许多教学建筑物，但是大部分是不用的。这一规划就是要在这个战略性的地点创造一个新的开敞区域，包括安置住在非法居住地的一些家庭（图11）。

图10　Paraná河岸的西班牙公园天际线

图11　城市中心的罗萨里奥大学

2. 南部峡谷

南部峡谷紧邻罗萨里奥大学中心，不久前这里还主要是非法居住地。倾斜的北端将改造成为公共空间，称为"新意大利公园"，这一计划的费用将会由意大利社区团体来支付。最后的城市重组项目尚处于发展阶段。

3. 南部地区公园（卡洛斯、西尔威、贝格尼斯博士公园）

南部地区公园位于Saladillo河的两岸，过去曾经是军营。其北部依靠罗萨里奥城市议会而南部则依靠Gobernador Galvez市议会。南部公园的功能包括运动和娱乐活动所需的设施，在属于罗萨里奥的部分，计划建立一条自行车道和一个KDT自行车赛道（已经完成）；在属于Villa Gobernador Galvez的部分，一个露营地和一个运动中心已经竣工。

七、结论

重建罗萨里奥境内Paraná河岸的工程,历经多届政府统治依然留存了下来,可归纳如下:

简单而动人的号召——恢复河滨原貌。

"将干预划分为阶段",为了让管理、规划和执行单位各自在不同的时期介入,有不同的角色及机会。

基于整个项目以及基于各领域、各方面多个单位的视角有所不同。这样,经济和生产活动并非完全与社会或城市分离而独立存在。港口工人的兴趣,与年轻人联系在一起,因为他们占用着公共空间;与带着孙子一起沿着河岸散步的老人们一致;与当地的房地产机构一致,因为他们预见了这片区域的发展机遇。城市为自身取得了新的面貌,并体现出本地区今后发展的远景。最重要的是,这个项目能带来很多收益。

一旦阿根廷的复杂情形有了改善并将经济危机抛诸脑后时,将再度制定条理清晰、协调平衡的规划方案。

参考书目

[1] Bragos O. Planes Urbanos, Espacio públicoy Proyectos de Ciudad: Revista A&P, 1996(11).

[2] Bragos O., Kingsland R. Organización Territorialy Nuevo Plan Urbano [Z]. Montréal, Canadian Instititue of Planners Conference "The City and its Region", 1999.

[3] Bragos O., Mateos A., Pontoni S., Vasallo O. Políticas Urbanasy Nuevos Roles de Ciudad Frente a Las Transformaciones Metropolitanas [M]//Bragos, O., Ribeiro, L. C. de Q. Territorios en Transformación. Rosario, Editorial Universidad Nacional de Rosario, 2002.

[4] Monteverde, R. Mejorando la Gestión Urbana / Unidad Temática de Planificación Estratégica de la Red de Mercociudades / Brasil – Argentina – Paraguay – Uruguay, [Z] 1999.

[5] Monteverde R. Ciudad Futura, Seminario Internacional de Metodologías de Planificación Estratégica[Z]. Rosario, Plan Estratégico Rosario, 2000.

[6] Plan Estratégico Rosario. Rosario, Municipalidad de Rosario, [Z]1999.

ROSARIO

Roberto Monteverde, Oscar Bragos

City Profile

Rosario is the only major Argentinean city that is not capital of a province. Located in the south of Santa Fe, its population is two and a half times higher than the one of the capital city and 30% of the province's population is concentrated there.

Even though the origin of the city dates from the 18th century, the building of the city, as we know it today started in the late 19th century. A prosperous trade centre gave birth to a city with high levels of infrastructural services that remain despite the deterioration caused by the economic crisis of the past few years.

The city's urbanisation pattern has always been of low density. This happens because the predominant type of house is the single house. The highest concentration of buildings with high levels of population density is only in the central area (where approximately 15% of the population lives). In contrast with other Latin-American cities, the centre is also characterized by the coexistence of a trade and financial centre and the residential area. Even though this peculiarity gives the central area more vitality, in the past few years the commercial centre has started showing some signs of deterioration and neglect.

Its inhabitants are mainly descendants of Spanish and Italian immigrants, but there are also some groups of Syrio-Lebanese people and Eastern European Jews. In the past few years, migrations from the north of the country have added new components to the ethnic diversity of the city, for example, the Tobas[①]. The quality of the public health care and the social policy attract poor and deprived sectors from the north of the country.

Small companies develop the industry. The most important branches are the textile, chemical, plastic, electric devices, furniture and food industries.

History of the City

Rosario has always been, and still is, one of Argentina's most important cities. Its economic development has always been related to the fact that it is located on the Paraná River. Policies to place it in the global market through the export of agricultural products, in addition to promotion of immigration and colonization, allowed Rosario to become the centre of a wide cereal-producing region, which had been almost totally unexploited until mid- 19th century. Until then, Rosario had been nothing but a small village with very few inhabitants near the boundaries of the natives' lands. The arrival of the railways, the construction of the port and the arrival of European immigrants quickly changed

---

① Natives settled mainly in Chaco and Formosa provinces in Argentina.

the life of this village and of its growing area of influence.

As a consequence, the city became the exporting port of the products from the Northern, Central and Eastern areas and, in 50 years approximately, saw its population grow several times. Since the city started developing, a conflict between landscape and production activities, between open areas and restricted zones came into being. The site of the conflict was the riverbank, where all the port and railway terminals can be found. The river and its bank have always been taken into account in every urban plan for Rosario, some proposals encouraged the relocation of those installations, leaving the zone free for public parks.

The first project for the expansion of Rosario appeared in 1873. In the project that seeks the growth of this city of 30000 inhabitants, the bank is considered one of the most important spots of the city. That is why the proposal includes an avenue on the ravines, like a promenade with viewpoints in each end. So, the ravines were considered from the very beginning a public recreational space.

Some years later, the French architect A. Bouvard, who headed the Plan d'Aménagément in 1910, reconsidered this idea. This time, the projected promenade would link different parks on the ravines. It was a greenbelt that separated the city from the port and that would visually connect the city and the river.

In the 1920s there was a movement that encouraged the beautification of the city. At the time, the city had over 250.000 inhabitants, its urbanised area had expanded northward, southward and westward, and the port was the second most important one in the country. New quality and quantity requirements appeared. It was demanded to turn Rosario into a modern city. That is, an efficient and beautiful city. One of the purposes of this local City Beautiful Movement, headed by traders and landowners was to get rid of port and railway infrastructures in the central area. The Sociedad de Ingenieros (Engineers Society) warned about the danger of these proposals that threatened the development of the city's most important activities. But those involved in the movement insisted and in order to achieve their goal they suggested using a new instrument: the regulating plan (master plan). Thus, in 1928 Rosario hired three local engineers to set up a plan that in 1935 was presented to the public opinion.

The aim of beautifying the city remained in the new plan — they wanted to move the port southward of the built area, leaving the central and northern banks free for parks, gardens and beaches. This plan was people's greatest ambition: to have the bank as a place for recreation, rest and sports. After 1935 and up to now, four new plans were presented and they all supported this idea of getting the bank for the people.

The 1953 plan was based on the redesigning of the railway network. After the war, the country's economic and political conditions had changed: industrialisation, nationalisation of railway companies, internal migrations. The city was part of all these changes; it reached the number of almost a million inhabitants and became part of an extensive urban conglomeration with nearby villages. Therefore, the idea was to free the bank of railway installations, keeping the port ones.

A new plan came out in 1968, with the same objective of gaining the bank for the people. It is basically the same idea as in 1935: to move the port from the north to the south and to convert the central bank into public parks.

A new plan came out in 1991. This time, keeping the old ideas, the riverbank was divided into 3 zones: the north, for sports and recreation; the centre, for recreation and cultural activities; the south, for port activities. But, in contrast to previous proposals, the port would remain in the north of the central area, modifying its condition of cereal-producing port into a terminal of containers. This was not an appropriate suggestion if we take into account the growth and consolidation of the city's urban area but it was corrected in the plan of the year 2000. The project of converting the bank into public space, which started 100 years ago, is nowadays beginning to be a reality.

Globalisation, Opportunities and Contradictions

The globalisation of the economy and alterations in state structure brought about significant changes in the territorial organisation of Santa Fe. Its south coast, where Rosario is located, is one of the zones where these changes became more evident. Today's territorial organisation is very different from the one there was ten years ago, and there is evidence that these changes will continue, which will make Rosario play a new role in the future regional scene.

Several factors combine to account for this situation. It can be attributed to the new orientation of public policies (end of the welfare state based on the principle of subsidiarity of the state with dismantling of state companies, privatization of services, transfer of lands left idle to the town council), of economy (industrial restructuring, service development, threatening control of the financial activity, unemployment growth), of the regional organisation (creation of the Central Region made up of the provinces of Cordoba, Santa Fe and Entre Ríos, insertion of Argentina to the Mercosur), of the social dynamics (internal migrations, increase of poverty and social exclusion). These processes, which characterize the changes that Rosario and its region are going through, refer specifically to the modernisation of the great infrastructures, industrial restructuring, new forms of urbanisation and the increasing worsening of social imbalance.

Since the coming into force of national laws that mark the end of state companies, there have been new projects involving the bank of the Paraná River. The end of the monopoly of the national state over the port activities marks the beginning of the formation of a network of ports along the Paraná which presents a great range of operations: export terminals of solid, mineral, oil and chemical products loads.

Besides, the region of Rosario and the metropolitan area will have, in a near future, a very different appearance regarding great infrastructures. The construction work in course will radically change the city insertion in the region. If the motorway that links Rosario and Cordoba, and the bridge that links Rosario and Victoria continue being built, Rosario will become a crossroads that will definitely complete the bi-oceanic corridor between the south of Brazil and Chile.

This promising works are contrasted with the economic crisis that has affected the industry in the metropolitan area. This stage of industrial restructuring is especially seen in a drastic shortage of employment, the closure of many businesses, the opening of new ones and the beginning of a process of internationalisation of the capital invested in manufacturing. The north metropolitan centre with the historical biggest industrialization of the zone has been the most affected with this process. In the mid 90s machinery and small chemical industries disappeared. This process accompanied some changes that had started in the mid 80s with a significant number of agro-industrial establishments (especially oil-producing ones). Their production is mainly exported. On the one hand, the south zone, which is smaller and traditionally oriented to meat processing activities, has gone through a simultaneous process of loss of small food and chemical industries and the emergence of big car parts industrial complex.

案 例

On the other hand, in the past few years Rosario's urban outskirts have recorded the existence of new process of land occupation. Related to new residence and commercial activities methods, these processes are important because they show a strong change in trends that had turned the outskirts into a receptacle of traditional ways of dividing the land into lots and of major projects of social housing. However, the old building methods have not disappeared and still coexist with these new ones.

Changes in the lifestyles of middle and upper class people, the recent developments in communications at every level, and the strategies of the real estate agents have led to new trends in the permanent housing submarket. In this way, people from Rosario have a progressive choice of neighbour villages as residential areas, in addition to the "private neighbourhoods" and districts of the west metropolitan centre. Furthermore, the new trade methods have developed in the past decade from specific needs of accessibility and localisation within the city, leading to a new morphological, functional and social map of the outskirts.

In the past few years, the process of marginalisation of vast groups of society has sped up. This is seen in the remarkable increase in the number of population that lives in the so-called "irregular settlements" and the fact that low class people have been pushed to neighbouring villages. This method of occupation of the urban zone increases year after year, and is characterised not only by the dispute over the use and ownership of the land but also (and most importantly) by the poor conditions of the whole environment.

Summing up, the changes brought about with the so-called "globalisation" take place in a contrasting atmosphere of uncertainties (the economy evolution), certainties (social exclusion) and challenges (adaptation of institutions to the changes, especially of the City Council authorities).

Urban Strategy

Until the last two decades of the 20th century, Rosario was always able to successfully agree with the major strategic decisions of the government. Therefore, when Argentina decided to join the world market as an agricultural and livestock centre, Rosario became the main cereal exporting port in the coast region. When, during international wartime, there was a process of replacement of imports, Rosario became one of the main metallurgical centres in the country, and a great number of industrial centres were established in the city and in the metropolitan area.

The turn of the century crisis, which affects the country, also affects Rosario. Consequently, the city's traditional profile starts to become blurred with no clear alternative regarding the new situation.

How can a new city profile be defined in an expanded territory? How can Rosario benefit from its location advantages regarding the great infrastructural works under construction? Local authorities have tried to answer these questions. They summoned the city's different protagonists of economic, social and cultural life to come up with a strategic plan. So, in 1996 they formed Rosario's Strategic Plan committee, which presented the final document in 1998.

The Strategic Plan's objective was, on the one hand, to define what the problems were and, principally, to identify the city's favourable conditions. On the other hand, it aimed at choosing and proposing projects and programmes that were essential for the definition of the city's new profile. The proposals are emblematic for transformation and become symbols of the city's re-foundation.

In outline, the strategy has certain projects that intend to take advantage of the city's new geo-strategic position, to improve the urban life quality and people's integration, to integrate the city with the river and promote the riverbank as a transformation spot. Among these projects, we find those related to regional links like the Rosario-Cordoba motorway, Rosario-Victoria bridge (both under construction) and the new metropolitan access roads; then those related to the quality of public spaces (region net of green zones, projects dealing with the restoration of the river bank), to the improvement of economic infrastructure (port and airport restoration), to the modernisation of municipal administration (decentralisation programme). Many of these projects have already been set in motion and are defining a new profile of the city. The changes in the river bank through the creation of parks in vast areas in addition to the formation of a technological centre, export consortiums and the regional development agency show this new profile and provide evidence of the close relation it has had, from its origin, with the river and the river bank.

Urban Issues in Local Strategy Management

According to these working criteria, the city has been building a strategy for its future, looking for a development project as well as for new management tools. Thus, urban projects have to be interpreted as a part of that general strategy.

From this strategy comes out the state policy modernisation, new social inclusion policies and urban projects that try to articulate themselves in a new and integrated model of city management.

The articulation suggested by the Strategic Plan lays on its own essence, because it is a common place where the initiatives -proposed by different actors- is gathered. The reconstruction of local dimension demands a place where agreements to plan city development can be established.

This global (holistic) approach guides the actions of those three columns of local city management: social plan, urban project and modernisation program. These three areas interact among them and with strategic plan. Social, economic and urban aspects are always present in every proposal. Projects are developed from this point of view; in many cases they have reached successfully their goals, while in others they have failed.

Urban Projects

Modernization and Municipal Decentralisation Programme

The aim of this programme is to make the administration more efficient, bringing the City Council (municipality) closer to the citizens, which implies an administrative reorganisation from the Centros Municipales de Distrito (District Municipal Centres), a redefinition of urban policies according to each district's peculiarities and a stimulation of the participation of the citizens in decision-making.

Nowadays, the city is divided into six districts and has three District Municipal Centres. On the other hand, these centres have organised meetings for citizen participation where the projects that deal with the improvement of life quality in each district are discussed, chosen and given approval.

Comprehensive Plan of Action in Irregular Settlements

The aim of this plan is to comprehensively solve the

problem of irregular settlements through interventions that include the building of new houses, urban restoration and social integration. In order to do so, the City Council has developed a specific programme called "Rosario Habitat", which has received financial support from BID (Banco Interamericano de Desarrollo) [①].

Rosario–Victoria Bridge

This work links the city of Rosario with Victoria (Entre Ríos province). It crosses the Paraná River and the islands of Alto Delta. It covers a distance of 59.5km (37miles) - 12.4km (7.8miles) of bridges and 47.1km (29.2miles) of embankment. It will be one of Mercosur's most important bi-oceanic corridors, an alternative to the tunnel that links Santa Fe with Paraná and the Zárate Brazo Largo complex in Buenos Aires. On the other hand, Rosario will be able to strengthen its role as a first rate agricultural centre and the main centre of interconnection with international markets. The work is currently in its final phase of construction.

Project of Riverbank Restoration

It is a series of projects considered as a whole and which are part of the transformation of the limit between the city and the river and the building of a new waterfront. Due to its nature and magnitude they are (together with the Rosario-Victoria Bridge) the symbol of a new city that is recovering from a deep and extensive economic crisis. These projects consist in the inclusion of new public spaces in the city and the building of new cultural and recreational equipment in the metropolitan area.

Rosario–Córdoba Motorway

The aim of it is to improve the present road connection between Córdoba (a city in the middle of the country) and Rosario (a city on the coast), becoming the main artery in the Central Region of the country. The first 17.65km (11 miles), which correspond to the Rosario-Roldán span, have already been completed.

Technological Park

The aim of this project is to create the right environment for the development of technology firms. For this purpose, the chosen place in the city should provide excellent environmental and infrastructural conditions and, at the same time, a direct link to research institutes. A Technology Centre has been formed- it groups computer companies and they are examining the different options to start the Technological Park.

Major Urban Project

River Bank Restoration Project

Once the building of the bridge on the river Paraná (joining Rosario and Victoria) is finished, it will open a new gate for the city to the east and it will incorporate a coast territory with particular landscapes and views.

The vision of the city that appears in the projects of restoration of the bank starts from an important reconsideration of the city's current bond to the river. It is expressed through a series of green zones and public

---

① Inter-American Development Bank.

buildings that will replace the old railway and port installations, which formed a rigid barrier between the river and the city. The projects give new value to a place that not long ago was not easily accessible and had an obsolete industrial image and they give new roles and meanings that symbolise the changes that the city expects.

The use relationship between the city and the river, created since the urban restoration of the Paraná's bank, has summoned the biggest amount of projects and interventions in the last decade. The target is to increase public spaces in the city. This new way of connecting the city to its river is shown in the different way of accessing the riverbank (which had been almost totally restricted not long ago); in the new uses (especially those connected to recreation and culture) and the construction of a new faÇade on the river. They all express the changes that the city is experiencing. The citizens' old ambition in the first decades of this century will now come into action- to free the river border from railway and port installations and to convert it in a free access recreational spot for general use.

By moving the port of Rosario to the south of the city and getting rid of the railway and port installations in the central bank, an area of 90 hectares approximately will be available for restoration and will surely be used for creating public spaces for equipment and recreation.

The restoration project is based on three principles that define the nature of the transformation in the sections in which the bank has been divided (north bank, central bank and south bank).

The first one refers to the regional dimension of the projects, taking into account, on the one hand, its functional diversity and, on the other hand, its metropolitan and regional scope. In fact, several situations that take place in the bank account for these two items:

> The development of tourism and recreation, especially on the north bank which includes a group of islands next to the city, belonging to Victoria Town Council, where in the past decade both the practice of water sports and the recreation on the islands have increased considerably.

> The development of metropolitan centrality on the central bank, particularly since new projects directed to commercial, cultural and recreational activities have been set in motion.

> The inter-municipal nature of some projects, especially the development of a great park in the south and a future new one at the head of the bridge in the north. All these added to the necessary agreements among the different town councils in order to stimulate a development of tourism on the island that respects the protection of the environment of the place.

> The promotion of the bank for residential development in neighbouring towns, both northward and southward of Rosario, which follows a trend and demands new road projects and the completion of the bank route in Rosario and the neighbouring towns.

The second principle points to the fact that the bank is public, which is accomplished by the creation of new green zones, equipped and accessible for everyone, on the central bank and southward, in Ciudad Universitaria (University Campus). The significant increase of the public park area is a policy from the part of the Town Council authorities to achieve democratization of city life.

Finally, the third principle refers to the assessment of different parts of the city related to the bank, especially the area of the south ravines, opposite the Port of Rosario. In this sector, the proposals point to an urban restoration which tends, on the one hand, to take advantage of the marvellous views to build new houses and, on the other hand, continue with the green zone network. These open areas will help mitigate the negative impact of the port activities in its surroundings.

North Riverbank Projects

Regional Park at the Head of the Rosario-Victoria Bridge

Rosario and Granadero Baigorria City Councils will carry out this project, since the biggest area of available land is located in the latter. The project intends, on the one hand, to arrange in order of importance the access to the bridge with a park and, on the other hand, to separate it from the residential areas nearby.

Special Plan for Environmental Organisation on the North Riverbank

The aim of the plan is to preserve the landscape characteristics of the slopes, defining the sites for recreational and gastronomic uses and setting the appropriate rules to protect the various existing viewpoints on the slopes.

Central Riverbank Projects

Scalabrini Ortiz Park

This site corresponds to an old place of repair shops of the first railway line in Rosario. Due to its location and size, the place was considered a barrier between downtown and northern neighbourhoods. Nowadays, a process of urban restoration has taken place as a consequence of disused railway lands given to the city. The project includes an area of major railway buildings (head office of the "Nuevo Central Argentino" company and a technical college); an area for the execution of commercial, cultural and residential projects; an area for a public park and an area for railway use. The aim is the creation of a new metropolitan centre and for it to rise as the northern gate to the centre of the city. The roadwork has already been completed while the investments for the building of a mall are momentarily frozen.

North Port and Scalabrini Ortiz Housing Park

This place is of high landscape value, since its geographic position offers great views of the islands, the central and north riverbank. Once the port installations are carried away there will be room for new activities and the Scalabrini Ortiz Housing Park. It is meant to become a metropolitan centre. The project consists in adding open spaces to the city keeping its great area condition of tertiary, cultural, residential and tourist-recreational activities.

Community Park (Parque de las Colectividades)

The silos, which are part of the city's architectural and urban patrimony, are in this site. It has recently become a park and now it will become public. Its shape will be the one presented by the winner of the Public Competition organised for this purpose. This new open space will be fitted with some restaurants on the edge of the slope.

Spain Park (North Extension)

This area has a series of disused railway installations on the top of the slopes, while the low riverbank has small buildings belonging to fishing clubs. The draft plans to keep the existing fishing clubs, foreseeing the development of cultural activities on the high sector. Besides, it includes the building of the Central District's headquarters office of the City Council Decentralisation and Modernisation Programme, restoring existing buildings, which belonged to the old railway administration.

Flag Memorial Park

It is the largest sector on the central riverbank and it is the city's most important civic and symbolic place and that has been historically associated with the central area. A great part of it is used for recreational purposes; other areas have nautical installations and restaurants, whereas those areas used for port activities have been long abandoned. It is an area in constant change as long as the lands that belong to state organisations are given to the City Council. Some of the latest works have been the Contemporary Expression Centre and the remodelling of the River Station (a restaurant and mall next to a passenger station). Since a definitive project has not been reached yet, the changes that are taking place nowadays follow a series of general directions given by City Council in 1996.

South Riverbank Projects

Rosario's University Centre

It is placed on the slope next to the Urquiza Park and opposite the river, separated from the port area by an avenue. It has a set of faculty buildings but there is a big portion of it left unused. The plan is to create a new open space included in this strategic spot of the city. The project includes the relocation of some families that live in an irregular settlement there.

South Ravines

It is located next to Rosario's University Centre. Not long ago it was full of irregular settlements. The slope's north end will be turned into a public space for the new Parque Italia. It is a project that will be paid for by the Italian community's institutions. The final project of urban reorganisation is still to be developed.

South Regional Park "Dr. Carlos Silvestre Begnis"

This area, which used to be a War Ministry camp, is located on both sides of the Saladillo stream. The north part depends on Rosario City Council and the south part to the Villa Gobernador Galvez City Council. Known as South Park, its project includes the construction of the necessary installations for sport and recreational activities. In the sector, which belongs to Rosario, the plan is to build a cycle track and a KDT circuit (already finished). In the sector, which belongs to Villa Gobernador Galvez, a campsite and a sports centre have already been built.

Conclusion

The project for the restoration of the Paraná River coast in Rosario has remained and survived different governments and it could be summed up in the following way:

> The strength of a simple yet moving idea: to recover the coast

> The possibility of "dividing the interventions into segments" in order to allow management, projection and execution units different from each other in periods, actors, opportunities.

> The base of the whole project and its units from multidisciplinary and multi-acting points of view. In this way, the economic and productive aspects are not completely separated from the social or urban ones. The interest on the part of the port workers is combined with the youths', who monopolise public spaces; the elders', who walk with their grandchildren along the riverbank parks and the real estate agents', who foresee opportunities in this area. The city gets a new look for itself and as a projection to the region. Many "benefit" from the project, which is a vital element.

That articulation and that balance will allow the project to be re-released when Argentina's complex situation improves and is left behind like a nightmare.

墨西哥城（墨西哥）

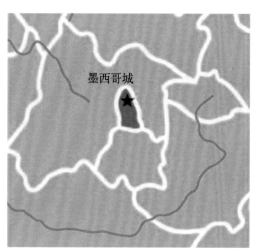

图 1　墨西哥城在墨西哥的区位图

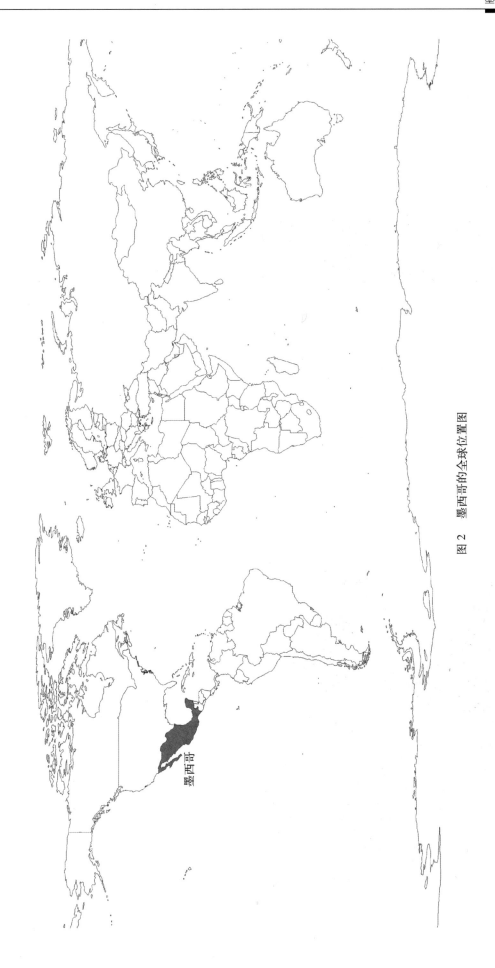

图 2　墨西哥的全球位置图

墨西哥城

亚历山德罗·帕莱翁

一、城市数据

（一）都市圈（MZ）

墨西哥市都市圈由两个不同的行政管理机构组成，一个是墨西哥市，另一个是由16个区与37个自治市所组成的墨西哥省（图3、图4）。

（二）城市化区域

面积：1800km^2。

人口（2000年）：18396677人。

人口年增长率：1.9%（1990～1995年）；1.4%（1995～2000年）。

人口密度：102.20人/hm^2。

出生于都市圈外的人口数及比例：6657721人（37.2%）。

18岁以上受过高等教育的人口数及比例：1926045（17.1%）。

缺少卫生设施的人口数及比例：8606862人（46.8%）。

EAP及比例（2000年）：7083965人（38.5%）。

第一产业：0.8%。

第二产业：25.9%。

第三产业：69.6%。

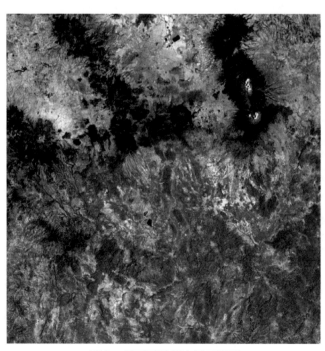

图3 墨西哥城都市区卫星图
（来源：墨西哥城市政府）

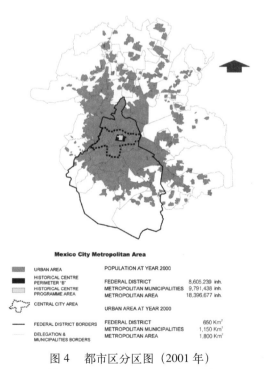

图4 都市区分区图（2001年）
（来源：A. Suarez Pareyon (ASP)）

收入（最低收入（MS），1MS = 115 美元 / 月）：

工作收入少于最低收入的人口数：575186 人（8.3%）。

工作收入介于 1～2 倍最低收入的人口数及比例：2370871 人（34.2%）。

工作收入介于 2～5 倍最低收入的人口数及比例：2342054 人（33.8%）。

私人小汽车拥有量及比例：1436945 辆（30.9%）。

（三）墨西哥城或联邦区

面积（1993 年）：670km²。

人口（2000 年）：8605239 人。

人口年增长率：0.3%（1995～2000 年）。

人口密度：128.4 人/hm²（2000 年人口数/1993 年土地面积）。

出生于联邦区以外的人口数及比例：1883831 人（21.9%）。

18 岁以上受过高等教育的人口数及比例：1236055 人（21.5%）。

缺少卫生设施的人口数及比例：3953017 人（45.9%）。

EAP 及比例（2000 年）：3582781 人（41.6%）。

第一产业：0.6%。

第二产业：21.2%。

第三产业：75.0%。

收入（最低收入（MS），1MS=115 美元/月）：

工作收入少于最低收入的人口数：301675 人（8.4%）。

工作收入介于 1～2 倍最低收入的人口数及比例：1141054 人（34.2%）。

工作收入介于 2～5 倍最低收入的人口数及比例：1187003 人（33.1%）。

住房数（2000 年）：

住房总数：2103752 套。

套均人数：4.1 人/户。

建设量：71660 套（3.4%）。

住房状况（2000 年）：

1 卧套房：345890 套（16.4%）。

1 室户：142333 套（6.8%）。

拥有独立卫生间：1921547 套（90.6%）。

设施齐全：2038157 套（96.9%）。

住房设施：

无：6336 套（0.3%）。

家电：

电视机：2037303 套（96.8%）。

冰箱：1801674 套（85.6%）。

电话线：1387907 套（66.0%）。

摩托车：816392 套（38.8%）。

基础设施与交通：

供水系统（1997 年）：35500L/s；353L/居民。

供电（1998 年）：12400GW/h。

汽油（1998 年）：125000 桶。

道路（1997 年）：8774km。

小汽车保有量（1999 年）：3265773 辆。

交通系统：

地铁年客流量（1999 年）：27 亿次。

墨西哥国际机场年客流量（2004 年）：2300 万人次。

固体垃圾（1999 年）：11561t/日。

教育：

小学在校学生（1998～1999 年）：1031789 人。

初中在校学生（1998～1999 年）：462812 人。

高中在校学生（1995 年）：329681 人。

高等院校在校学生（1997 年）：298314 人。

大学及技术学院：

公共大学：4 所。

研究中心：12 个。

私人院校：90 所。

卫生设施：

医院和健康机构（1999 年）：1051 家（15782 个床位）。

休闲：

公园：105 个。

绿地（1998 年）：4452hm² （人均 5.3m²）。

博物馆：88 个。

图 5　中心区：向北看
（来源：A. Suárez Pareyón）

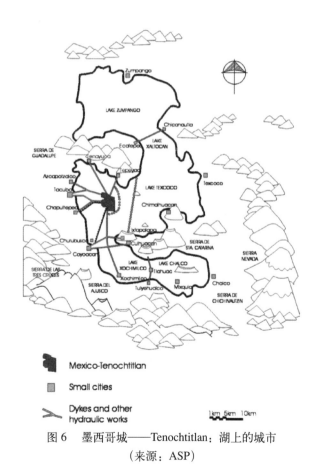

图 6　墨西哥城——Tenochtitlan：湖上的城市
（来源：ASP）

图书馆：222 个。

剧场：104 个。

电影院：269 家。

商业（1999～2000 年）：

公共市场：312 个。

街道市场：1051 个。

流动市场：59 个。

超级市场：346 个。

（来源：Cenvi's Research Programme；Observatorio Urbano Ciudad de Mexico OCIM.2000 年国家普查信息。Consejo Nacional de Poblacion CONAPO y La Ciudad de Mexico Hoy，Bases Para un Diagnostico (2000)）

二、城市概况

经历了一个世纪墨西哥城才从一个只有适度人口的城市变为世界上最大的城市之一。1900 年它的人口是 34 万，到 2000 年达到 1820 万[①]。墨西哥城是一个巨大的城市聚合体，包括两个管理和体制上的区域：联邦政府管辖区和墨西哥州政府管辖区。联邦政府位于联邦管辖区内，而墨西哥州政府只是这个国家的 32 个州中的一个，这两个区各有其自身的地方政府。联邦辖区从 1997 年 12 月首次由市长选举制度代替了传统的直接由总统任命制，从而引起了一系列的变化。相反，在墨西哥州及其辖区则不同，从 19 世纪开始，无论是地方官员还是州府官员都是通过民主选举而产生的。联邦区由 16 个选区组成，并管理他们各自的地域，墨西哥州辖区也一样。现在，这两个区的人口也几乎是相同的。

墨西哥城的区域面积超过 1300km²，包括联邦辖区的 16 个选区及墨西哥州辖区的 37 个自治市。最近还从 Hidalgo 州划过来一个自治市（图 5～图 7）。

---

① 2000 年的人口普查结果。

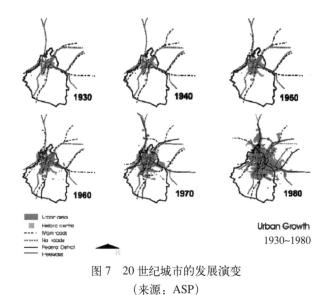

图7 20世纪城市的发展演变
（来源：ASP）

（一）墨西哥城的经济重要性

城市人口和占地面积的生长一部分原因是农村向都市迁移，还有一个原因是历史上经济活动、功能及政治因素所导致的空间集聚。在1940～1970年之间进口替代政策导致墨西哥以工业生产为基础来带动经济增长。很显然，由于鼓励产业向墨西哥城集聚的政策，它成为经济增长最快的城市，这些加速了城市经济活动的发展和集中。

在1970～1980年期间，经济增长伴随着通货膨胀和国家石油资源的开采和出口而持续。另外一个国家经济发展的原因是在北部边境地区设置了Maquiladores，这意味着墨西哥城的企业开始为全国及地方消费而生产。

众所周知，1980～1988年是经济危机时期，进口替代模式的失败导致墨西哥城的工业生产急剧下跌，也因为此，墨西哥城失去了在当时国内众多城市中的领导地位。1988年以后，墨西哥城完全接受了新自由经济的模式，通过市场控制工资和汇率水平。另外，国家还开放了贸易竞争，实行财政改革，实施对国有企业的私有化，解除了对金融体系的管制，并都与政府脱钩。墨西哥城的工业继续下降，但是服务业却开始呈现发展趋势。

尽管这样，墨西哥城在整个国家的经济中仍然起着很重要的作用，1998年全国比较中，墨西哥城区占了国内总产值的32%，但是联邦区只占国内总产值的22%。

（二）墨西哥城：矛盾的城市

20世纪末，墨西哥城在社会、经济和文化上充满着矛盾的现象。对于贫穷的研究表明，在墨西哥城，62.2%的人是穷人，35.5%的人是极端贫穷。1990年，市中心周围55%的地区住的是穷人，他们违章占有土地建造违章住房。这里大约有160万户居民，总人口达到820万人。这些状况也反映了过去的50年时间里，城市发展中的一些不平等特征。

三、城市历史

墨西哥—特诺齐提特兰城是在1325年由墨西哥人，也就是国际上所认为的阿兹台克人，在今天墨西哥国家所在的中央高原上的一个湖中心建立的，是通过一些自然形成的小岛经过Chinampas的使用后构成的，利用了湖畔的美索亚美立亚文化中心Plateau的水利技术。一些早期西班牙观察员把这个系统叫做"飘逸的花园"。墨西哥城是一个岛屿城市，通过在堤坝上建造的大路连接到陆地，并调节控制湖中的水。与祖先的方法相同的是，城市呈现出以十字架形状为基础的长方形布局，十字架的线条都有着各自的宗教意义。一个主要仪式场所建在十字架两条线的交点处，其南侧和西侧则建有总督的宫殿和公共事务管理部门。在远处，中心地区周围的住房被组织在比较小的街区里，这些部分共同构成了城市的四个都市区的一部分。

在残暴和血淋淋的争斗以后，西班牙人征服者1524年在宗教中心和墨西哥市的废墟上重建了墨西哥城。新建的城市在过去西班牙式城市结构的基础上进行建造，但是中心地区的开发并未达到他们前人的水平。在西班牙占领的三个世纪期间，先前的西班牙式复杂排水设施遭到毁坏或由于缺乏使用而失修，因此城市经常有洪水泛滥。这就是 Virrey 政府决定抽干墨西哥湖修建新的排放系统，并导致墨西哥湖生态系统破坏的主要原因。

19 世纪有超过一半的时间是在战争中度过的，开始是为了独立而反抗西班牙帝国主义，而后是为了抵抗欧洲人和北美人的入侵，后来又有自由党和保守党之间交战。尽管有着极大的人力和经济损失，墨西哥城在这些冲突过程中维护了它的重要性和统治地位。19 世纪 50 年代以后发生的这些政治和经济变动，打破了过去的做法并且重新定义了城市的未来。一方面，自由主义者在击败保守主义者后为促进城市改革制定了新规则，允许城市通过 Colonos 的新邻里计划来进行扩展。另一方面，国家放开从发达国家引进资金和技术，导致了铁路的产生和蒸汽电力在工业生产上开始使用。墨西哥在 19 世纪的最后 20 年和 20 世纪早期由 Porfirio Diaz 将军统治，在他的独裁统治下，墨西哥就体验了类似于今天广为人知的"全球化"时期，国家开始了工业化进程，同时，天然原材料大量被发达国家开采，尤其是美国。这样，在一个贫困的农业化社会中，开始接受来自法国的时尚与生活方式，向英国学习技术手段等，墨西哥城实现了引人注目的发展。

现代化，是 20 世纪的头十年墨西哥城最有代表性的一个词，大规模公共项目的建设，提供了基本的城市服务和运输供应，社会设备、公共建筑和一些重要的大厦建设完毕，城市空间品质也得到了改善。城市规模充分扩张，现代化带来了面貌改善和工业化。主要经济和行政活动位于城市的古老区域，大多数住房也是这样。在 1910～1920 年十年多的时间里，这个过程由于墨西哥革命爆发而停止，整个城市的建设活动几乎停滞，只有少数几个投资者敢于冒险投资大楼与城市化项目，当地及国家政府为了生存根本无暇顾及公共项目的建设。战后又持续了十年的不确定时期，直到和平来临的 20 世纪 30 年代开始，城市发展才在工业化的推动下，再度开始，这时墨西哥城的居民只有一百万人。

在 1930～1950 年间，墨西哥城的中心是城市主要公共及私有建筑及房地产投资的主要地点，建设行业也得到了快速增长，第一幢摩天大楼在这期间建成，居住建筑也得到了很大的发展。在这一时期，城市中心区内的旧街区成了新移民的落脚地。1950 年以后，移民则集聚在城市周边地区，促使了都会区范围的扩大（图 8、图 9）。

直到 20 世纪 60 年代前，墨西哥城中心曾经由于公共投资的作用而具有很强的经济活力，而后，投资者发现新的区域更有吸引力，在某些情况下，经济活动从市中心迁往这些新的区域。结果，城市原有中心地区的退化加速，现代化的办公区和商业活动被吸引到条件更好的地点。

图 8　都市圈边缘的 Nezahvalcoyolt 市
（来源：墨西哥市政府）

20世纪后半叶墨西哥市的发展模式是一个综合了社会、人口、经济、政治和法律因素的复杂网络所作用的结果。农村人口由于失去了赖以生存的土地或其他资源，于是向城市，尤其是墨西哥城迁移，为了获得就业机会及改善自身的生活状况。由于城市规范的土地市场无法提供给他们足够的土地、服务及住房，非正规的土地、服务及住房市场随之产生。战后的土地改革带来了土地的社会所有制，很多土地脱离了规划控制及政府管辖，变成为没有基础设施、服务和社会配套的巨大区域。结果，政府采取了"对于正在发生的事情睁一只眼闭一只眼，放任自流"的态度，直接影响到20世纪最后30年里地方政府的态度（图10、图11）。

1980年墨西哥城都市区的人口超过1370万，1985年9月19日城市遭遇了里氏8.4级地震，给市中心带来了惨重的影响，即使在二十几年后的今天，灾害的后果仍然能看到。政府和整个社会做了大量的重建工作，在四年间建造了6万套住宅。

一方面，墨西哥的经济发展模式在20世纪80～90年代期间开始转变，起始于智利学校的自由主义思潮开始影响到墨西哥政府，而20世纪的最后十年间全球化的巨大浪潮则对墨西哥城产生了最巨大的影响。未来的城市包括中心区西南方向空置的和未充分利用的地方，譬如建筑材料商店，被改造成与全球化相协调的新用途，包括跨国企业总部、旅馆、高收入住宅、商业中心、娱乐中心，以及为领导者提供新的发展模式培训课程的教育中心。另一方面，都市区获得了持续增长，虽然与早先十年相比增长速度更慢，部分原因是由于土地的非正规利用以及由于贫困而缺乏城市发展的服务设施。

四、全球化，机遇和挑战

墨西哥政府努力满足了国际货币基金组织的要求，使北美洲自由贸易协定得以在加拿大、美国和墨西哥三国之间签署。然而，不平均的经济实力和协定

图9　卫星城的大厦，雕塑的设计师为 Barragan
（来源：A. Suarez Pareyon）

图10　鸟瞰的历史中心
（来源：墨西哥市的市景）

图11　墨西哥城的大教堂
（来源：A. Suarez Pareyón）

的条款实际上有利于美国，协定只是人们有关经济可以一体化的想象，主要受益人是那些以较低的成本在墨西哥生产、为美国及世界市场提供产品的跨国企业，他们甚至不需要墨西哥的多少投入。一旦美国经济出现衰退时，情况就会变得非常复杂。

在20世纪90年代期间，全球化对于墨西哥城的影响在房地产行业最为突出，1994年的金融危机使市场活动降至冰点，所有工程都停顿下来，建筑业成为最受打击的行业。虽然整个市场在最近三年里有了小幅回升，但距离重新恢复活力依然遥不可及。

与新自由主义坚持减少国家干预自由市场的思想相悖，在全球化的问题上政府依然采取干预的态度，通过明确的政策引导，以吸引投资者有兴趣参与墨西哥城的开发建设活动。显然，城市发展规划体系需要现代化，并且在城市公共事务管理层面上不能屈居次要地位。

1997年，重新修订了联邦区和大都市区的城市规划体系，自1976年第一次通过之后就再没有改动的城市开发法也在同年修订完毕。1987年制定的城市发展通用计划也作了类似修改。由于缺乏一种规划机制，城市不得不实行新的城市"巨型"项目，即吸引私人投资参与一些特殊项目，而整个墨西哥城都市区缺乏为整体发展而设立的规划体系，有鉴于此，墨西哥州进行了有关城市群协调发展的规划研究。

在1996~1997年间联邦区修订了城市开发法，1997年修订了城市一般开发程序，位于联邦区的16个选区也修订了各自的开发方案。1998~2000年间，31个城市战略区的局部开发计划被通过，其中24个在联邦区的议会获得通过。

五、城市战略

从1998年开始，联邦区新政府承担了31个战略区的局部开发计划，占地13000hm^{2}并位于所有16个选区，涉及120万居民。局部开发计划要求被开发地域的规划师与当地组织和机构间直接沟通，也包括对住宅、区域组织和政府的作用。这是第一次由各方共同来研究人的居住问题，而不是通过政府出面组织。大学规划系、城市规划部门和私人专家与政府部门密切联系，共同在这块特殊的土地上推动城市发展进程，并且经常与社区代表和地方组织一起沟通。

2000年举行了联邦区的州长竞选，并且首先采取的一项措施是禁止在城市的周边建造新建筑。建筑许可证只发给那些位于联邦区中心地区的新建筑，或者是对老建筑的修复。这一行动的目的是避免城市在联邦区保护区内过度成长，同时，促使基础设施在人口流失的中心地区再度得到利用。这个措施是非常有争议的并且它对都市发展的作用仍然是未知数。

（一）有关新机场选址的争论

政府制定的最有争议的决策之一是墨西哥城新国际机场的选址。虽然根据全球化进程的需要，而现有设施容量不够并且进行扩建的可能性不大，所有的部门都认同了新建一个机场的必要性，然而，新机场的选址仍然很有争议。两个选址位于联邦行政区外的大都市范围内：一个在墨西哥州，与已干涸的Texcoco湖相连的巨大延伸区域内；另一个在Hidalgo州内，由Tizayuca市政当局管辖内的农村地区。

不同选址的优缺点在激烈的讨论协商中，来自各州政府的政治家都在竞相吸引机场建设可能带来的巨大投资。虽然双方政府都雇用了最好的技术与谈判专家，但是辩论现在变得非常透明，政治家们及其技术代表的观点直接展现在公众面前。

Texcoco这一选址有一些优势，因为目前这块基地尚未使用，并且积满了水。基地距离城市中心35km，占地14000hm^2。机场建筑的建设将提供32000个工作职位，以及在现有机场的20000个职位以外新增7000个职位。这一选址被批评的主要原因是它在生态保护上很有价值，同时，这一基地的建设中将会遇到一些麻烦。从城市发展角度来看，它将对现有城市结构有所调整，同时却无法保证能够解决现有以及未来可能遇到的一些问题。

选址在 Tizayuca 意味着利用农村土地，它位于离城市中心 65km 处，占地 4000hm²。它将会提供 5000 个直接就业岗位以及 15000 个间接工作岗位。这个选址方案还要求建造一个运输系统，并且改进现有的区域间联系，或者建造一个新的交通网络。从城市发展的角度来看，Tizayuca 选址看起来能够支持都市区在将来 20 年里的发展。

最近，联邦行政区的总督倾向于支持 Tizayuca 方案，因为 Texcoco 方案可能存在影响城市持续供水的问题，特别是排水系统方面。同时，Texcoco 方案建成后将关闭位于联邦行政区内现有的国际机场，因为两者相距不远，而这将会使城市蒙受巨大的经济损失。Tizayuca 方案则不会威胁到城市水力系统的稳定性，并且现有的机场仍然可以继续使用。

最后的决策完全是技术决定的，联邦政府下定决心并启动了 4550hm² 的土地征用，并未考虑任何社会或政治敏感性问题。被征用的土地是由几个不同的农会组织自治管理的 ejido 土地。因此，好几个小镇因此到农业法庭起诉该项土地征用。联邦政府制定的补偿标准为 7 比索 /m²（合 0.65 美元 /m²），但立即遭到农会组织的断然拒绝，称它是一个政府迫害。农业法庭作出了有利于农会的裁定，其导致的政治冲突迫使联邦政府不得不让步，新机场计划因此而取消。

联邦政府官员随后发表的声明中说，新机场并非十分必要，现有机场的过度饱和可以通过在机场附近新建一个航站楼以及将部分航空公司的航线分散到中心区域其他机场来缓解。其中一个备选机场是 Toluca 国际机场，它在 2 年内就成为一些低成本航空公司的营运基地，所有的私人航空公司也转移到这里。墨西哥城国际机场的 2 号航站楼始建于 2002 年，总建筑面积 22 万 m²，总投资 4.4 亿美元。联邦政府机场管理委员会确信 2 号航站楼完成后能够满足客运量的需要。2004 年墨西哥城国际机场的客流量为 2300 万人次，当新航站楼启用后，总客流量预计将达到每年 3200 万人次（图 12、图 13）。

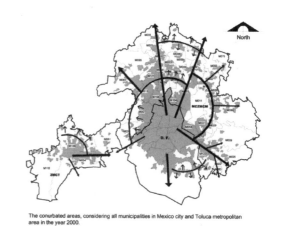

图 12　1990～2000 年间墨西哥城都市圈的扩张模型
（来源：作者）

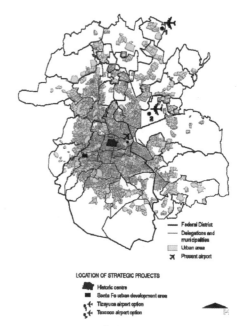

图 13　2000 年战略项目的分布图
（来源：墨西哥市政府）

六、城市大规模开发项目

（一）圣达菲项目

圣达菲项目被公认为是最能在全球化阶段时期体现墨西哥市发展的一个项目。项目开始于 1988 年国家政府更迭后 Salinas 总统执政期间，由于其新自由主义模式和经济开放政策的使用而闻名于世。圣达菲位于城市西部的一个群山脚下，项目覆盖了数百公顷的土地，那里原先是在20世纪时用来堆放建筑材料的，而露天的矿产资源正被周围日益增长的城市所包围。这块土地的所有权是属于联邦区政府的，改善后由政府机构的代理Servimet负责销售。圣达菲是联邦行政区保留下来的唯一土地，用地广阔，有着足够的空间来进行多功能的开发，例如跨国公司办公楼、商业中心、私立大学、豪华住宅、娱乐设施以及绿化景观等。它的重要意味着它是这一完整的经济发展长廊的最后一个环节（图14）。

（二）历史中心的再生

墨西哥城具有历史意义的中心传统上被称作为"中心"，这意味着它是这座城市里最为重要的地方。不管如何变革、如何衰退、房屋倒塌、人口流失，以及经济和功能中心变迁，这个中心对于这个城市甚至国家的重要性是不会改变的。它集合了最多的历史性建筑物，这些建筑物至今仍在使用。这里有1800座列入保护名单的建筑，而且这一数字随着新名单的制订而继续上升。对这些建筑和城市空间的保存和修复始于1934年，在1987年，联合国教科文组织宣布了这座历史中心对于人类的重要性（图14、图15）。

历史中心面积为9.7km^2，1995年的人口超过了18万，这几乎达到了联邦行政区总人口的1%。

直到1950年，历史中心由于其在经济、政治、行政活动上的集中性，以及作为一个文化标志，一直是城市最重要的区域。在那个时候，它汇集了大量的人口，一些区域的人口密度也超过了每公顷600人。然而，土地使用功能的改变以及建筑物缺乏维护、住房短缺、1985年的地震影响、部分经济与政府活动分散到其他地方，这些都加速了人口的流失。1970～1995年之间，历史中心在流失了11.86万居民后每天仍有120万流动人口，并在用地的公共及私有利用方面存在着利益冲突。这些历史中心多样的生活对于城市里那些占有着废弃与破败建筑的弱势人口仍然具有吸引力。

从1998年开始，联邦行政区政府注意到历史中心的战略价值，并开始重视为其制定发展政策。在地方政府的协作下，代表着区域里居民的组织、商业机构、非政府组织，以及各个大学，一起为历史中心的复兴而努力，目标是对当地的整体复兴。

与此同时，历史中心信托基金会负责推进和管理针对中心的投资，并为历史中心的复兴与综合开发制定一个战略规划。

联邦行政区内的城市发展与住房管理部，得到CENVI这个非政府人居组织的技术支持，为历史中心制定局部城市开发计划。另有两家私营咨询机构为历史中心的另两个区域制定城市开发计划，一个是La Alameda，另一个是La Merced。

图14 建筑师Gonzalez de Leon设计的圣达菲大厦
（来源：A. Suarez Pareyon）

图15 Polanco酒店区的企业大楼
（摄影：Alejandro Suarez Payeron）

联邦行政区社会发展部（SDMFD）与荷兰的开发援助组织（NOVIB）共同支持了 CENVI 发起的历史中心整体重建项目。

当联邦行政区政府、联邦政府和私人投资者一致同意对那些参与历史中心建筑重建工作的投资者和设立的公司进行财政优惠后，2001 年 8、9、10 月历史中心的重建工作有了进展，这项协议于 2001 年 10 月 8 日正式由总统发布。

（三）新住房发展计划

为了解都市区内住房建设大规模增长的显著性，有必要先了解联邦政府住房政策的调整，特别是住房金融体系的改变。半个世纪前，政府设立了住房基金（FOVI）以支持公共住房项目建设，向商业银行提供贷款担保。1972 年住房共同基金设立，通过 INFONAVIT 项目为正式产业工人、通过 FOVISSSTE 项目为政府雇员、通过 FONHAPO 项目为失业工人提供帮助。

墨西哥银行于 1982 年实施国有化，1990 年开始私有化，1993 年再次成为商业银行。经济上的自由化导致了住房政策的调整，联邦、州及市机构间制定了协调机制，以加强他们与私人及社会机构的联系，其主要目标是开放机会以满足住房需求，提供住房开发商和建造商进入原有管制市场的机会，也满足劳动者通过一定的住房融资基金实现住房需求的愿望。这一改变需要整合原来因限定于不同目标群体（INFONAVIT、FOVISSSTE、FONHAPO 以及使用商业银行资金的 FOVI 项目）而分隔的房地产市场。

从 1995 年起，住房战略是基于提供购买能力的思想并侧重于长期住房发展，为此，政府的角色限于推动、协调、与住房相关的各方面订立协议，以及支持住房的生产、融资及取得。同时，政府力求减少技术、管理、法律障碍，以减少住房建设的时间和成本。

到 20 世纪末，联邦区内保留的土地因大型住房项目而使用殆尽，土地价格高企使得大型开发企业开发廉价住宅的想法难以为继，因此他们开始在都市区范围内寻找农业用地，并与地方政府达成一致用于大型住房开发。

地方法律法规体系在土地利用规划等方面存在大量的漏洞，导致建成的大规模住房项目缺少足够的基础设施与服务。近几年，都市区北部和东部这些比较贫穷的地区都批准建设了大量的住房项目，这对原有居民和新迁入这些住房的居民带来严重的问题，包括饮用水、排水管道、教育和公共卫生设施等问题以及交通问题。

一个特别需要指出的问题是，从联邦区到相邻的墨西哥州移民导致了市区人口大量增加。人口及住房普查报告显示 2000~2005 年有 25.8 万移民从联邦区移民至墨西哥城都市区。

同期，在都市区范围内针对缴纳住房基金（主要是 INFONAVIT 和 FOVISSSTE）的工人阶层需要新建了 16.8 万套住房单位。2006 年，廉价住房市场呈现出饱和的迹象，供应开始超过需求。

（四）墨西哥城市峡谷郊区铁路

墨西哥城市峡谷郊区铁路项目是国家政府与联邦区政府及墨西哥州政府共同推动的几个项目之一，通常被称为"郊区火车"。

50 年前，国家政府决定采用基于小汽车的土地交通政策，未对公众作充分的宣传就开始减少铁路的数量。当整个铁路体系瓦解后，铁路线路及设施被出售给跨国企业专门用于货物运输。

由于都市区内出现的严重客运交通问题，国家及地方政府最近决定恢复年久失修的郊区交通项目。项目将在历史中心边缘重建铁路终点站 Buenavista 并重修废弃的线路投入使用。第一条线路是从 Buenavista 至城市北侧的一个集中区 Cuautitlan，线路总长 27km，共设 5 个车站。铁路由电力驱动，设在联邦区内的两个车站与城市地铁系统相连。初期的服务将满足每日

图 16　铁路车站及中部地区的工业园区
（来源：墨西哥市政府）

32 万人次客流的需要，老的 Buenavista 车站将建成商业中心，并连接到地铁 8 号线以及沿 Insurgentes 大街行驶的有轨公交车。

2005 年，西班牙 CAF 建设公司（Construccionesy Auxiliares de Ferrocarriles S. A.）赢得了该系统建造和未来 30 年运营权的招标。初期建设及第一条线路的运营投资为 5.93 亿美元，另加 1.15 亿美元的意外准备金。郊区火车将于 2007 年年末投入运营，联邦政府运输与交通部还宣布了在都市区东部建设另外两条线路的许可（图 16）。

七、结论

大墨西哥城市（包括联邦区和都市区）与其他拉丁美洲的大都市区一样，面临着贫困背景下人口和城市快速发展而带来的问题。半个世纪时间里，政府经历了难以估量的增长规模与复杂性，规划与设计则严重不足，资金、技术、政治资源都难以满足全球化背景下这个城市社会的庞大需要。

21 世纪伊始，墨西哥城就站在发展的十字路口，现在的联邦区政府持左翼政治与思想立场，有机会通过选择保持公平而促进发展的模式而改变城市。然而，如果考虑到墨西哥州政府与国家政府不同的政治立场，上述目标则不是那么容易实现。这些差异在近期的总统选举和市政府选举中都有所体现。2007 年年初，必要的沟通与寻求城市一致发展的努力，使得官方建立联合与有效的都市协调机制的努力开始产生实际效果。

参考书目

[1] Cenvi's Research Programme：Observatorio Urbano Ciudad de Mexico OCIM [Z].

[2] Coulomb René. El Centro Histórico de la Ciudad de México al Fin del Segundo Milenio，pp.530-537，Coordinado por Gustavo Garza [Z].

[3] Datos del Observatorio Urbano de Ciudad de México Basados en el Censo de Población de 2000 [Z].

[4] Damien Araceño. "Pobreza Urbana en la Ciudad de México al Fin del Segundo Milenio，pp.297-302，Coordinado Sobrino, Jaime. "Participación Económica en el Siglo XX, en la Ciudad de México al Fin del Segundo Milenio"，pp.162-167，Coordinado por Gustavo Garza por Gustavo Garza [Z].

[5] Instituto Nacional de Estadisticas, Geografia e Informatica INEGI, Press Information January 29th 2001 [Z].

[6] Information Processed of National Census 2000, Consejo Nacional de Poblacion CONAPO [Z].

[7] Programa General de Desarrollo Urbano del Distrito Federal 2003 [Z].

[8] Suárez Pareyón, Alejandro. Encenarios Socioeconómicos y Espaciales de la Zona Metropolitana de la Ciudad de México [Z]. Revista El Mercado de Valores.

[9] Suárez Pareyón, Alejandro. "El Centro Histórico de la Ciudad de México. Presente y Futuro, en los Centros Vivos, Alternativas de Hábitat Popular en los Centros Antiguos de Iberoamérica", Coordinado por Rosendo Mesias y Alejandro Suárez Pareyón [Z].

[10] La Ciudad de Mexico Hoy, Bases Para un Diagnostico (2000) [Z].

MEXICO CITY

Alejandro Suarez Pareyon

Mexico City is one of the biggest metropolis in the world. In 1900 its population was 345000 habitants and in 2000 it had 18.2 million inhabitants. Mexico City is an agglomeration, which contains urban areas from two administrative and political demarcations, one being the Federal District (FD) and the other being the State of Mexico. Mexico's Federal Government is located in the FD and the second demarcation is one of the country's 32 Federal States. Both demarcations have their own local governments. Since 1997 the FD held elections for Governor after a long tradition of its Mayor being appointed directly by the President. The FD is made up of 16 delegations that govern and administer their respective territories. These are similar to the municipalities in the State of Mexico. At present the population of the Metropolitan Zone located in the Federal District and in the State of Mexico is almost the same. It occupies 1300 square kilometres, including the 16 delegations of the FD and 37 municipalities in the State of Mexico.

The demographic and physical growth of the city has been partly due to migration and also for the historic tendency of the spatial concentration of economic activities, functions and political decisions. Between 1940 and 1970 the import substitution policy led to Mexico's economic development on the basis of industrial production. Clearly Mexico City was the most dynamic area for industrial growth as its industrial promotion policies favoured the location of industries in Mexico City. This increased the growth and concentration of economic activity in the city.

During the decade 1970-1980, economic growth was accompanied by inflation and the exploitation of the country's oil wealth. Another major element of the country's development was the setting up of 'maquiladores' on the northern frontier (massive serial commodity production at very low wages). This meant that industries in Mexico City began to direct their production to national and local consumption.

The exhaustion of the import substitution model was initiated during the period 1980 and 1988 and was reflected in a deep economic crisis. Since 1988 Mexico fully adopted the neoliberal model. Additionally, the country was opened up to trade competition, fiscal reforms, privatization of public enterprises, deregulating the financial system, and cutting the government apparatus. Industry continued to decline in Mexico City and service sector activities grow. At the end of the 20th century Mexico City was full of contrasts where different expressions of social, economic and cultural development could be identified. 62.2% of the total population could be defined as poor, and 35.5% of the total population could be defined as being extremely poor.[1] In 1990, 55% of the human settlements

---

[1] Damien, Araceño, Pobreza Urbana, en *La Ciudad de México al Fin del Segundo Milenio*, pp297-302, coordinado por Gustavo Garza[Z].

that surrounded the central area of the city used irregular land occupation and building methods of the poor. These popular neighbourhoods had a population of around 8.2 million inhabitants.

Historic Background

The city Mexico-Tenochtitlan was founded in 1325 by the Aztecs, in the centre of a lake, which corresponds to today's Mexico. The city Mexico-Tenochtitlan was constructed on some naturally-occurring small islands which were extended by use of *chinampas*, which are parcels of floating agricultural land achieved through the application of hydraulic technology known to the lakeside cultures of the Mesoamerican central plateau. Mexico-Tenochtitlan was an insular city linked to firm land by avenues constructed over dykes that held-in and regulated the water in the lake. The city was planned to form a rectangle based on the shape of a cross where both of its lines had ritual significance. After fierce and bloody battles, in 1524 the Spanish conquerors re-founded the City of Mexico over the ruins of the religious centre and the Mexica city. The new Spanish city was rigorously planned on the structure of the prehispanic city, and was developed over the central part without even reaching the same dimensions as its ancient predecessor. Over the three centuries of Spanish domination, the complex system of prehispanic hydraulic infrastructure was destroyed or fell into disrepair due to lack of use, and the city was flooding continouosly. This is why the Virrey governments decided to drain the Lake of Mexico, they instead constructed an elaborate system which led to the destruction of the lake ecosystem.

There were wars for more than half of the 19th century, beginning with the Independence struggle against Spanish imperialism and later with the defence of the country against invasions from Europe and North America, and additionally confrontations between liberals and conservatives. Despite enormous human and economic loss, Mexico City maintained its importance and hegemony throughout these conflicts. Political and economic changes were taking place from the middle of the century, which broke with the past and defined a new future for the city. On one hand, the liberal victory over conservative forces led to new laws that promoted Urban Reform, allowing the expansion of the city through the creation of new neighbourhoods called *colonos* in Mexico. On the other hand, the opening-up of the country to investments and technological progress from industrialised countries led to the coming of the railway. During the last two decades of the 19th century and the first of the 20th century, Mexico was governed by General Porfirio Diaz, and under his dictatorship Mexico experienced a first period similar to what is known today as Globalization. The country began to industrialise and, at the same time, developed countries, particularly the United States, exploited its raw materials. The fashions and life styles used came from France, the technology from Britain and the remarkable development of the city was also achieved by an impoverished rural society.

Modernity was the catchword in the city during the first decade of the 20th century and large-scale public works were undertaken, the provision of basic urban services and transport, social equipment, public buildings and some prestige buildings were constructed, and urban spaces were improved. The city was in full expansion; modernity had brought improvements and industrialisation. The main economic and administrative activity were located in the ancient city, and so was most housing. Between 1910 and 1920 the Mexican Revolution stopped this process, and construction in the city came almost to a standstill. Between 1930 and 1950 the centre of Mexico was the key location for the city's main public and private buildings and real estate investments. As a result, the construction industry grew significantly. From 1950 onwards the migratory flow was to the city periphery, which contributed to the physical expansion of the large metropolitan area.

The centre of Mexico City was intensely economically active, largely due to public investments, until the 1960s, however from then onwards, new areas were attractive to private investment and economic activities found more attractive sites in other areas. In some cases, economic activities left the centre of the city for these locations. As a result urban deterioration in the most popular parts of the central area accelerated and modern offices and commercial activities were attracted to better locations.

During the second half of the 20th century Mexico's rural population migrated to the cities, and particularly Mexico City, in search of employment and improved living conditions. Informal and irregular land was a result of the insufficient supply of land for housing, services or housing and their availability on the formal and regular markets. The Agrarian Reform created socially owned land, which left huge tracts of land outside the control of the local government. These turned into huge areas for housing without infrastructure. As a consequence the unwritten policy of *turning a blind eye to what is happening and let things take their course*, has had long-term implications for local government during the last thirty years of the century. In 1980 the population in the Mexico City metropolitan area was more than 13.7 million inhabitants. In 1985, an earthquake measuring 8.4 points on the Richter scale shook the city and the central area was the hardest hit. The consequences of the disaster can still be seen today.

Mexico's economic development changed during the 1980s and 1990s under the influence of the liberal Chilean school. The city of the future included the southwest of the central area where vacant and underused lots have been exploited for new uses. These included the head offices of transnational companies, hotels, high-income housing, commercial centres, recreation centres and education centres with courses designed for the leaders of the new development model. In contrast, the metropolitan area continued to grow although at a lower rate than in previous decades, and was still partly based on irregular occupation of land and lack of urban development associated with urban poverty.

The North American Free Trade Agreement signed between Canada, United States and Mexico did not lead to expected results. The main beneficiaries were the transnational companies that had located in Mexico to produce goods at low cost for US markets. Globalisation in Mexico City during the 1990s was felt particularly in the real estate investment sector, and until the financial "crash" of 1994 markets were frozen. A small upturn has been observed during the last few years although market reactivation is still far from being achieved. Contrary to the neoliberal philosophy of reducing State intervention in the free market, this has not been wholly followed. It became clear that the urban development planning system required modernisation and could no longer occupy a secondary position in the city's public administration hierarchy.

Strategic Urban Planning

From 1998 the new FD Government undertook 31 Partial Urban Development Programmes, covering together almost 13 thousand hectares and located in all the 16 Delegations. The population attended by this planning effort is 1.2 million habitants. The elaboration of the Partial Plans required a direct relationship between the professional planners and the civil and community organisations of the small territories being planned. New elections were held for the FD Governor in 2000 and one of the first measures to be taken was prohibition of new construction in the delegations on the periphery of the city. Planning permission would only be granted for new construction, or the rehabilitation of old construction, in the FD Central Area. The objective of the action was to avoid urban FD growth in conservation areas, and at the same time, allow the reuse of infrastructure in the Central Area, which had lost population. This measure has been very controversial and its impact on urban development is still unknown.

The Santa Fe Project

The Santa Fe Project is considered the most representative in the period of Globalisation. It began in the 1988, period known for the rigorous application of the neoliberal model and the opening-up of the economy. Santa Fe is located in the west of the city on the lower reaches of a group of mountains. Land is the property of the Government of the Federal District and was improved and sold through the government agency SERVIMET. Santa Fe is the only territorial reserve in the Federal District with sufficient space for a multi-functional development, including offices for trans national companies, commercial centres, private universities, high cost housing, spaces for recreation activities and green areas. Its strategic location means that it is the last development of a whole economic and development corridor.

Large Urban Project: Regeneration of the Historic Centre

The Historic Centre of Mexico City is the most important place in the city, and despite the transformations, the deterioration, the loss of housing, and the population expulsion, the loss of "centrality" conditions, the Historic Centre is still important for the city and for the country. It contains the largest concentration of historic buildings, which are in use. There are 1800 buildings that have been considered for conservation. Protection of these buildings and urban spaces began in 1934 and in 1987 UNESCO declared the importance of the Historic Centre to humanity.

The Historic Centre is 9.7 square kilometres and in 1995 its population was more than 180 thousand habitants. Until 1950, the Historic Centre was the most important space in the city due to its concentration of economic, political and administrative activities, and as a cultural symbol. At that time densities were 600 habitants per hectare. However, land use changed, the physical deterioration of the buildings, the loss of housing, the 1985 earthquakes, the decentralization of some economic and government activities led to accelerated population loss. However, the Centre still received a floating population of about 1.2 million people every day, generating conflicts of interests over the use and occupation of public and private land. The intense life in the Historic Centre means that it is attractive for the city's vulnerable population who occupy the abandoned, or ruined, buildings.

From 1998 the Government began to recognise the strategic value of the Historic Centre and gave importance to its development policy. In a co-ordinated effort between the government of the city, the organisations that represent the inhabitants of the zone, the business sector, non-governmental organisations, and universities work began to recuperate the Historic Centre. Integrated regeneration was their goal. At the same time, the Historic Centre Trust Fund was responsible for promoting and administering some of the investments to the Centre and, in addition, it developed a Strategic Plan for the Regeneration and Integral Development of the Historic Centre.

The Urban Development and Housing Ministry for the Federal District, with technical assistance from CENVI, which is a human settlements non government organisation, developed the Partial Urban Development Programme for the Historic Centre. Two other private consultancy agencies developed Partial Urban Development Programmes for two other areas of the Historic Centre, one being La Alameda, and the other, La Merced.

上海（中国）

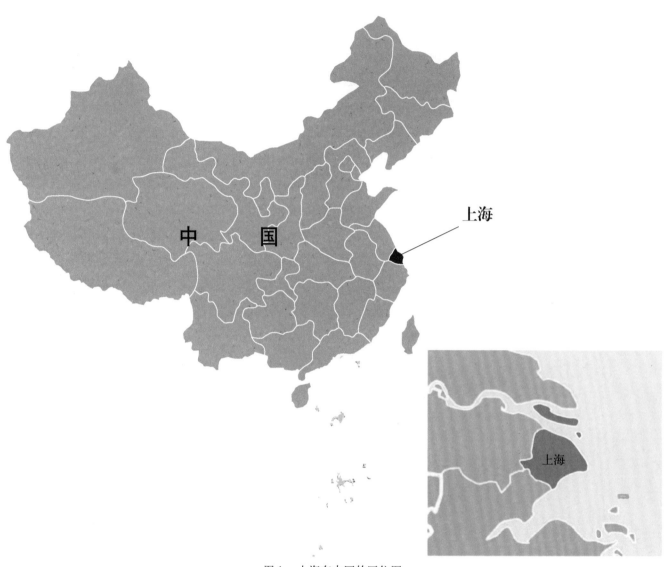

图1　上海在中国的区位图

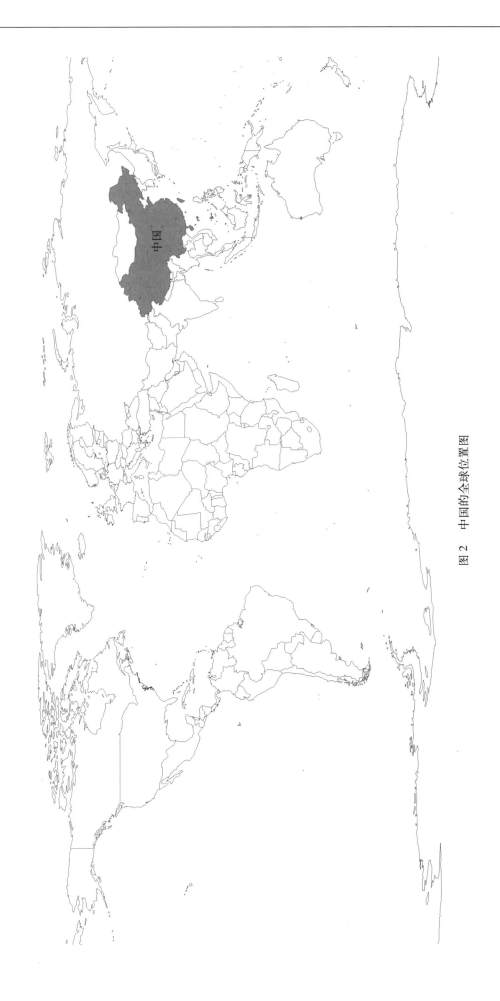

图 2　中国的全球位置图

上 海

陈雅薇　胡　昊

一、城市数据

(一) 土地与人口

土地面积：6340.5km²。

行政区划：18区、1县。

总人口（户籍人口/常住人口）：1391.04万人/1888.5万人。

户籍人口年自然增长率：-0.75‰。

户籍人口中非农业人口比例：87.5%。

人口增长：

1990年，1283万人。

2000年，1321万人。

2004年，1352万人。

人口密度：2978人/km²。

中心城区人口密度：13597人/km²。

(二) 经济指标

GDP：13698.2亿元（2005.6亿美元）。

人均GDP：72534元。

GDP年增长率：9.7%。

外国直接投资项目数：3748个。

外国直接投资合同金额：171.1亿美元。

实际到位外国直接投资金额：100.84亿美元。

进出口总额：3221.4亿美元。

出口：1693.5亿美元。

进口：1527.9亿美元。

税收总额：2382.3亿元（348.8亿美元）。

城镇登记失业率：4.2%。

(三) 住房、基础设施及其他社会指标

家庭总数：506.6万户。

平均家庭人数：2.7人/户。

人均可支配收入：

城区：26675元/年。

郊区：11385元/年。

市区人均居住面积：16.9m²。

农村居民人均居住面积：62.3m²。

城市基础设施投资：1733.2亿元。

城区排水管线总长度：9208km。

城区燃气供应量：28.4亿m³/年。

公共绿地面积：34256hm²。

(四) 教育、健康及休闲

新出生人口期望寿命：81.3岁。

每万人拥有医生数：27人。

医院及医疗中心数：2809所。

医院病床数：97800张。

（来源：上海市统计年鉴，2008）

注：人民币汇率为1美元=6.83元人民币。

二、城市概况

上海,是中国的四个直辖市之一,直接隶属于中央政府管理,人口超过1300万,地域面积为6340.5km^2,它位于长江和长江三角洲地区的联结点,长江三角洲地区是世界上经济发展最迅速的地区之一。上海有18个区县(黄浦、卢湾、徐汇、长宁、静安、普陀、闸北、虹口、杨浦、浦东、闵行、嘉定、宝山、松江、青浦、南汇、奉贤17个区以及崇明县)。上海是中国最大的城市和最重要的经济、贸易中心和工业基地,也是中国最主要的港口,开放的沿海城市,著名的历史和文化城市,在中国享有重要和特殊的经济和社会地位,自20世纪以来吸引了越来越多的全球目光。随着上海建设国际化大都市步伐的加快,它已经为自己勾画了一幅成为国际经济、金融、贸易及航运中心的美好蓝图(图3)。

三、城市历史

上海原本叫华亭郡,建于751年,大致位于今天的松江区,从今天的地理位置来讲,其北部延伸到现在的虹口区,东至下沙,南部到杭州湾的海畔。在991年的时候,上海镇建立。在1260~1274年间,上海发展成为一个重要的贸易港,1292年,当时的中央政府在这一地区建立了上海县,这被广泛地认为是上海这座城市的开埠标志。20世纪20年代,外部力量使上海迅速成为一个世界知名的城市。如今,全球化又一次为上海重新成为国际化大都市发挥了巨大和深入的作用。

上海最初是一个小渔村,几个世纪前也仅仅是一个小城镇。而就是在19世纪后半期,由于中国在一系列对外战争中的惨败,上海不得不被迫对外开放,而成为了一个半殖民地城市,一些地区被割让给外国列强成为租界。外国侵略者在侵略的同时也带来了先进

图3 上海的区位

技术和现代文明。在1949年新中国成立之前,上海是亚洲地区除东京之外最为高度发展的城市。但是新中国成立之后,上海和新中国一样被隔离在世界之外,接下去的30年时间里仅仅维持了中国大陆一个重要工业基地的地位。20世纪80年代开始,中国采取了对外开放政策,上海又重新恢复了它与国际社会的紧密联系。特别是从20世纪90年代以来,全球化,主要是以外商直接投资和改造的形式,使上海又重新成为了一个国际化的大都市。今天,上海已经形成了一道独特的风景线,既保存了历史上西方的建筑特色,又融入了近年来上海新发展的现代化摩天建筑群(图4)。

近年来上海呈现在世界面前的令人印象最深刻的就是它快速的城市化和大规模的城市基础设施建设。因为新中国成立后政府公共投资捉襟见肘,以及忽视城市公共基础设施建设的建设方针,上海市中心人口密度极高,居民生存环境低下。过去的十年中,上海市政府在城市化和重建高标准的基础设施上花了大力气,大量的住宅群已经建造起来了,居民区和在市中心的工厂也被重新规划,尤其是占地约500km^{2}的浦东新区已经为上海的发展提供了新的空间。

上海在国内外都享有盛誉,不仅因为它是一个繁荣的大都市,更因为它具有丰富的近现代人文资源。

案 例

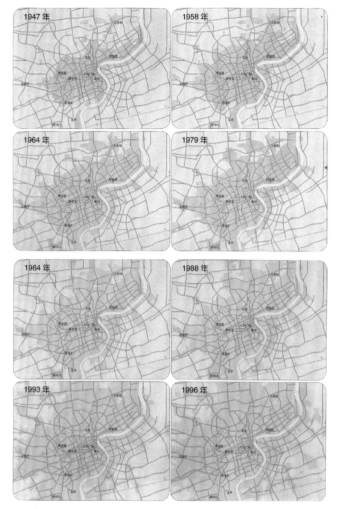

图4　上海城区的扩展（1947～1996年）

图5　外滩夜景

图6　浦东陆家嘴

图7　20世纪初的外滩

图8　20世纪30年代的外滩（对面是浦东）

其"海派"文化在中国也是十分著名的，如"海派"建筑所具有的华美屋顶等特色（图5、图6）。

16世纪（明朝中期），上海成为全国纺织品手工业的中心。1685年，上海建立了中国第一个海关。当时上海已经是一个拥有20万人口的城市了。19世纪中期的鸦片战争以后，上海变成了一个半殖民地半封建城市。同时，它进入了现代贸易和工业发展的阶段，对中国内陆扮演着主要贸易港口和出入口的角色。一些国外列强在上海建立了租界地。在20世纪20年代，租界地发展到了它的鼎盛时期，拥有超过200万的人口，并且使上海成为当时世界上最著名的国际性城市之一。1936年，上海被列为全球第7大城市和亚洲最现代的城市（图7、图8）。

1949年5月27日，中国人民解放军解放了上海，上海也开始巨变。从20世纪50年代一直到70年代，上海一直是中国最大的工业基地。在中央政府的紧密控制下，上海是当时国家财政收入的主要贡献者，在20世纪70年代大约占到全国总量的25%。当中国在1978年开始改革开放政策以来，上海在社会和经济发展上都经历了显著的进步。持续快速的发展大大增加了城市的综合实力和人均国民生产总值。2003年，上海的人均国民生产总值已经达到了46517元（5618美元），这表明上海已经达到了世界中等发达国家的平均水平，同时也是中国大陆最发达的城市。现在上海正在为成为世界经济、金融、贸易、航运中心而不断努力。

四、全球化，机遇和挑战

全球化在上海发展成为国际化大都市的过程中起到了戏剧性的作用。19世纪后半叶，上海不得不对西方列强实行开放政策。1853年，上海取代广州，成为中国第一大贸易城市。重商主义进入上海和中国现代高等学府的建立使上海成为中国东部的金融和文化中心。外国殖民者在上海建立了租界地，并自行负责管理。法国殖民者建立了电力网络、排污设备，以及与当时欧洲大部分城市相同的电车系统。那个时期上海发展成为了亚洲最大的城市。

新中国成立后的上海一直处在中央政府的严格控制之下，所以在20世纪80年代的时候，上海的发展一直落后于中国南方珠江三角洲的一些城市。只有在1990年以后，中央政府宣布开发建设浦东新区时，上海的城市发展才获得了极大的空间。全球化在上海建设成为世界级城市的过程中，又一次起到了主要作用。全球化是一个复杂的过程，主要表现在两个方面，一个是跨国投资，一个是人才流动。全球化最重要的形式是跨国直接投资，包括合资、合作和独资。截至2003年年末，上海已签订外商直接投资合同投资额744.37亿美元（其中2003年110.64亿美元），实际吸收资金额462.65亿美元（2003年58.5亿美元），外商投资合同项目32061个（2003年4321个）。另一方面，全球化在很多情况下，采用的是一种软件流动的方式，包括先进的技术、管理方式、西方文明、价值体系等。例如，越来越多的国外城市规划设计师和建筑师开始参与到上海城市建设工程的竞标活动中。

全球化在上海的社会和经济发展中起到了直接和巨大的作用。首先，全球化以外商投资的形式给上海带来了众多资金。上海的许多企业在全球市场中缺乏竞争力，面临重组、雇佣劳动力减少等问题，甚至濒临倒闭。上海市政府无法拿出充足的资金来发展城市经济。在这样的情况下，国外的投资就成了一种重要的资源。因此，吸引外商投资成为上海市政府优先考虑的议程。事实上，来自世界银行和亚洲开发银行的资金，对上海的很多重要的建设项目提供了资助。在过去的十年中，如果没有国外资金，上海不可能完成那么多的城市建设工程。其次，先进的技术和管理方式随着外商在中国或上海的投资而被引进。上海采取了优先政策，划定实施范围，来鼓励对高科技工程的投资。第三，这种作用也同样体现在了对城市风格的改造中。上海正在逐步成为一座越来越现代化的城市。整座城市都在进行旧城改造，新的摩天大厦正在旧市区建立。现在，超过8层的高层建筑数已经达到了4226幢，占地面积为7.41亿m^2。其中超过20层的有1690幢。第四个作用就是城市的管理方式正在发生改变，尽管这是无形当中发生的。外资是一种超出政府直接控制范围的资源。另外，当外商面临来自国内外的激烈竞争时，上海市政府已经开始调整它自身的管理模式，来满足外商发展中对公共服务的合理需要。

虽然全球化给上海带来了巨大的国际影响力和极大的发展机会，但还是有一些问题不能忽视，而且这些问题很可能会产生一些矛盾。随着国外投资和跨国公司的纷纷进入，地方企业将会面临越来越激烈的竞争。当上海的发展主要依赖于国外资金的时候，人们也会很自然地怀疑这样的一种发展模式是不是可靠。当矛盾爆发的时候，类似于1997年亚洲金融危机的一系列问题是不是还会出现。尽管上海现在的人均国民生产总值已经超过了5600美元，预计2007年将会达到7500美元，但是还必须清楚地看到，整个中国的人均国民生产总值还仅仅是800美元。中国的其他地区和上海之间的贫富距离正在拉大。同时，在上海，富人和穷人之间的距离也在快速拉大。如果这个问题得不到合理的解决，整体社会将很难谈得上和谐与稳定。

五、城市战略

上海的发展战略将分为两个步骤：长远的发展目标和中期的五年计划。

上海已经为社会和经济的发展制定了新的中期与长期战略计划，力图尽快将上海建设成为世界经济、金融、贸易、航运中心，以及具有社会主义特色的国际化大都市。在这一长期规划下，2010年的目标是：

(1) 初步形成一个经济发达、综合实力强大的世界级大都市；

(2) 将都市的空间布局最佳化；

(3) 建立初步的城市规划；

(4) 形成初步的城市化；

(5) 全方位参与国际经济竞争；

(6) 将市场机制引入社会主义经济中；

(7) 追求社会、经济发展和环境保护之间的平衡。

战略规划的目标是在2020年把上海建设成为一个世界级大城市。未来的5~10年对上海来说将是一个至关重要的时期。首先，上海需要努力保持人均国民生产总值的快速增长，在2007年之前达到7500美元。其次，上海需要继续实施城市基础建设的大规模改造，其中关键的工程包括深水港建设、国际航空港建设、高速铁路系统建设和地铁网络建设。第三，上海将会继续加强软件建设，比如国际化、城市管理、信息系统、市场机制等。同时通过和江苏省的南京、苏州、无锡、常州、扬州、镇江、南通和泰州，浙江省的杭州、宁波、嘉兴、绍兴、舟山的合作，上海将致力于建立世界上除了纽约、墨西哥城、东京、巴黎—阿姆斯特丹、伦敦—曼彻斯特城市带以外的第6个大都市发展带。

紧随中央政府的计划，上海市政府也制订了新的五年计划，来规划中期发展的目标，为2001~2005年的社会和经济发展作了指导方针、目标和主要发展战略上的安排。

六、城市大规模开发项目

（一）浦东新区规划

浦东新区的开发开始于1990年。从那时起，它就已经在新的城市系统的开发、工业升级、开放政策的试点等方面扮演了先锋角色。这一特定区域是城市旧区以东毗邻的三角地带，面积为533km²，人口为163万。浦东由四个重要的开发区组成：陆家嘴金融贸易区、外高桥保税区、张江高科技开发园区和金桥出口工业加工区。陆家嘴地区规划面积28km²，为上海提供了一个新的中央商务区（图9、图10）。

过去，由于缺少桥梁和隧道等基础设施，浦东隔绝于上海老城区的经济发展之外，尽管与后者只是隔了一条黄浦江。1990年，浦东的产值只有60.24亿元，

图9　1990年浦东新区开发前的外滩（对面是浦东）

图10　2004年的外滩（对面是浦东）

而到了2003年，这个数字已经增长到了1507.44亿元，增长了25倍多，而且年平均增长率高达28.1%。到2010年，一个由金融、贸易和高新技术组成的工业发展系统和技术创新系统将在浦东新区建立。一个经过科学规划，有着良好的环境和完善设施的新城将被建立起来。经济和社会平衡发展，人与自然协调生存，物质和精神共同进步的新的发展模式将在这里找到良好的发展空间。通过工业升级、功能开发和系统改革，浦东的竞争力将会得到显著提升，它对于长江三角洲经济发展的积极影响力，也将会在长江流域和其他地区得到进一步加强和提升。

（二）"一城九镇"规划

为了疏解城市中心区人口高密度发展带来的各种问题，上海希望将中心区人口吸引到城市近郊生活，用一个由中心城市、新城、中心城镇和新城镇构成的崭新的城镇体系来取代以前的单一核心的中心城区城市结构。作为一个试行计划，"一城九镇"规划将松江发展成为新兴城市，将安亭、罗店、朱家角、枫泾、浦江、高桥、周浦、奉城和堡镇建设成为中心城镇。在规划中，这些城镇将被设计成不同的国际风格，甚至有很多按国际标准设计的办公楼与住宅。

不过这些不同风格是否能够与当地已有的建筑与环境风格相互融合，目前还是一个有争议的话题。在开发时间上，这些城镇的同时建设，使得相互间对于建设资金和大规模城市基础设施，特别是公共交通设施的竞争趋于白热化（表1）。

（三）上海"新天地"改造工程

上海新天地是上海具有浓厚"海派"风格的都市旅游景点。它的前身是上海近代建筑的标志之一——破旧的上海石库门居住区。改造之后，上海新天地被创新地注入了诸多时尚的商业元素，变成了一个集餐饮、购物、娱乐等功能于一身的国际化休闲、文化、娱乐中心，成了一个具有国际知名度的聚会场所。同时，新天地项目的建设已经成为中国房地产区域改造的经典案例，其成功的策划、规划、设计、建设以及经营模式长时间以来一直被众多城市管理者及城市开发商们青睐，很多人都想"移植"其成功的旧城改建模式。

新天地的建设是太平桥改造项目的第一期，坐落在上海市中心的卢湾区，繁华的淮海路商业街的南侧。1996年，香港瑞安集团受卢湾区政府委托，邀请了国际著名的美国SOM公司对整个地区做了规划，规划占地52hm^2，总建筑面积约130万 m^2，目标是将这一片旧城区改造成一个标志性的现代化综合区，以配合上海发展成为国际大都市的需求。

按照规划，太平桥人工湖绿地成为本地区的中心，西面为新天地，东面将改造成一个综合性的购物娱乐商业中心，南面是配套完善的翠湖天地高级住宅小区，北面规划为企业天地甲级办公楼区。

以先改善环境来提升该地区的品质和价值，带动整个地区的房地产开发。开发新天地，为该地区营造了一个高品位的人文环境，开发人工湖绿地，为该地区创造了一个优美的生态环境，提升了52hm^{2}土地的品质，再进行房地产开发，经济效益、社会效益、环境效益三赢，这个旧城改建的新理念已成为房地产开

案例

上海"一城九镇"项目简介

表 1

城镇名称	风格	设计师/设计事务所	简介
松江	英式	英国阿特金斯	占地 22.4km²，东边为景观河道，南边建设行政中心，北边建设大学城以及交通枢纽站。城西南 1km² 用地将建成英国特色高档住宅区，老城则保留历史文化名镇的风貌，形成"一城两貌"
安亭	德式	阿尔伯特·施贝尔/德国 ASP	规划用地 4.9km²，居住人口 8 万人，绿化覆盖率将达到 70%，人均拥有绿色空间超过 600m³，湖泊总面积将达到 80 m²。依托上海国际汽车城，以汽车产业、房地产和旅游三大产业为支撑，力争用 5 年左右时间建成高品位、高起点、特色鲜明的现代化城镇
罗店	北欧	—	规划面积 6.8km²，将建成具有典型北欧城镇居住特色、交通特色和环境特色的现代化生态新城镇。其中建设用地 3.4km²，城市森林 3.4km²，规划人口 3 万人。新建镇的整体规划包括有占地 1.2km² 的核心风貌区和占地 2.26km² 的 36 洞国际级高尔夫比赛球场
朱家角	本土水乡古镇风貌，又有现代城镇的格调	—	规划常住人口 5.5 万人，其中 0.68km² 为古镇区，保持明清风貌，3.5km² 体现具有中国特色的现代风格。规划以"风、光、水、绿"作为空间元素，发挥其生态平衡优势，计划在 2007 年将以旅游为未来的定位基本实现。与其他"一城九镇"不同的是，这是唯一一定位中国传统特色风貌的古镇，修复、保护老镇区与开发新城区两者并重
枫泾	北美式	美国 NBA 规划公司	作为紧邻浙江的门户古镇，枫泾镇将以服装机械为产业支撑，以商贸流通、旅游休闲能、建设具有北美特色风貌的现代化生态城镇。镇区规划面积为 8km²，人口 7.5 万人，分为 3.0km² 的新镇功区，2.1km² 的老镇区，1.4km² 的商贸区以及 1.5km² 的门户公园 4 大板块
浦江	意式	卡纳第/意大利格里高蒂公司	以 300m×300m 的方格街区作为基础，自行车道穿插两条东西走向的道路风格。区内穿插两条东西走向的步行街和自行车道。区内中心位置设计了一个公园，以及两条南北向的马路。人行道、自行车道设计为广场，其形态各不同，边缘设计小商店和服务性行业。浦江镇的各个交叉处设计为广场，通过严格的管理，以企业为主体实施市场化运作。整个城镇建设周期约 10 年，投资将达 50 亿元人民币
高桥	荷式	Kuiper Compagnons 和 TKA 设计公司	北部的现代风貌区由老镇脱胎而来，保留部分浓郁中国传统风貌建筑，将具有现代荷兰特色的布局和建筑造型，规划建设高级居住社区，传统风貌与现代风貌与人工环境、现自然景观与自然景观的有机融合。保留建设具有荷兰特色风貌建筑，将具现自然景观的有机融合
周浦	—	—	—
奉城	西班牙式	—	城镇布局突出体现西班牙地中海城市的风格。镇区周围还将建设大面积生态林、世纪森林公园、现代农业园区等，并构筑 60km² 的绿色生态景观。到 2020 年，奉城镇占地面积将达 16.08km²，总人口 7.2 万人，成为一座西班牙建筑文化与江南古镇特色风貌和谐统一的特色生态城镇
堡镇	—	—	—

152

图 11　上海新天地

图 13　新天地鸟瞰

图 12　上海新天地的发展规划

图 14　新天地的街景

发的新亮点，也反映了城市建设的新思路（图 11 ~ 图 14）。

1. 太平桥人工湖绿地

人工湖绿地占地 4.4 万 m²，水面面积为 1.2 万 m²，绿地下面的地下停车场可提供 200 多个停车位。这是由上海市政府、卢湾区政府、瑞安集团三方共同投资的，由瑞安集团负责物业管理。整个项目耗资近 10 亿元，2001 年 1 月开工建设，2001 年 6 月竣工，如今，太平桥人工湖绿地已经成为市中心一个独具特色的景观空间。

2. 上海新天地

上海新天地是一个具有上海历史文化风貌的都市旅游景区，占地 3 万 m²。它以东西方文化融合、历史与现代对话为基调，以上海传统的石库门建筑旧区为基础，配合充满现代感的新建筑群，改造成具有国际水准的集餐饮、商业、文化、娱乐为一体的综合性时尚休闲步行区。

新天地分为南里、北里两大部分。以石库门建筑为主的北里，集合了来自世界各地风情的餐厅、咖啡室、酒吧、精品商店等，充分展现了新天地的国际元素。以现代化建筑为主的南里，除了临湖的 88 新天地酒店式服务公寓外，更有一座总面积达 2.5 万 m² 的购物、娱乐、休闲中心，是适合时尚青年一族的休闲娱乐场所。

新天地已发展成为上海的时尚新地标及品位的象征，成为一个张望上海昨天、今天和明天的窗口。

3. 翠湖天地高级住宅小区

翠湖天地位于人工湖南侧，是整体规划中占地面积最大的项目，总建筑面积 68 万 m²。坐拥充满海派风情的上海新天地和宁静雅致的太平桥绿地，翠湖天地试图为现代精英提供一处理想的家园。

4. 企业天地甲级办公楼区

企业天地坐落在太平桥人工湖北侧,办公楼沿湖而建,形成长1.2km的湖滨建筑,总建筑面积逾50万m^2,将兴建多幢甲级办公楼、酒店、商场及其他配套设施。一期总建筑面积7.8万m^2,包括两座办公大楼,目前已成为众多跨国企业追逐的办公区。企业天地是淮海中路东段核心商务区的扩展,目标是营造一流的商圈,发展成为跨国公司总部所在地。届时,独具品位的办公楼将与新天地的石库门建筑交相辉映,完美协调,为国际都会增添一座优雅尊贵的经典建筑。

(四)黄浦江改造规划

早些年,上海就为黄浦江沿岸的改造制定了一个雄心勃勃的计划。黄浦江是上海的标志,而且为上海提供了诸如外滩之类的旅游景观。长达114km的黄浦江发源于上海西部的淀山湖,穿过市中心,最后流入长江。400m宽的黄浦江还在上海的水陆交通中起到了重要的作用。黄浦江流经市中心的部分,西岸就是外滩。外滩被看做是上海的标志。在长度1km的外滩,分布着各式各样的西方建筑,有的甚至建于20世纪初。外滩更是因为融合了各种不同的建筑风格,而被人称作是"万国建筑博览会"。黄浦江的东岸就是浦东新区。陆家嘴金融贸易区正对着外滩,摩天大楼林立。东方明珠电视塔是亚洲第一高度,金茂大厦是世界上第三高楼。陆家嘴除了集中了众多高楼大厦之外,还是国外金融机构和国内生产市场的集中地(图15~图18)。

上海市政府估算,完成黄浦江两岸综合改造项目大概需要花费1000亿元人民币,折合120亿美元。它被认为是继浦东新区之后的另一个里程碑式的战略工程,预计在十年内完成。这项工程将涉及22.6km²,长达20km的黄浦江沿岸地区。在听取了六个重要的国际性组织和设计公司的意见之后,上海市制定了初步计划,将黄浦江两岸综合改造工程分成四个区域分

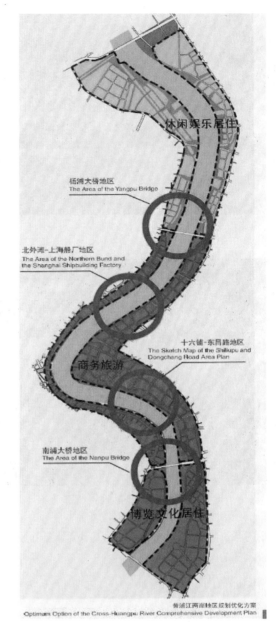

图15 黄浦江两岸综合开发功能规划分区
(来源:黄浦江两岸综合开发办公室)

图16 北外滩总体规划
(来源:RTKL)

图17　北外滩详细规划效果图
（来源：RTKL）

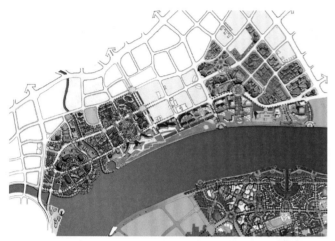

图18　北外滩详细规划总平面
（来源：RTKL）

块进行，分别是杨浦大桥地区、北外滩地区、十六铺—东昌地区和南浦大桥地区。具体计划见表2。

上海黄浦江两岸综合改造工程分区　　表2

区域	面积（km²）	功能规划
杨浦大桥地区	2	以住宅为特色
北外滩地区	1.15	购物和观光
十六铺—东昌地区	0.35	文化遗迹和观光
南浦大桥地区	3.43	现代建筑和2010年世博会场所

这些位置现在大多是废弃的仓库、码头和工厂，将来将被建成现代办公楼、商务楼和住宅区以及休闲设施。

1. 起因

上海市政府发展这一区域的起因很简单——黄浦江河道的功能和沿岸设施已经不能满足上海发展的需要了。上海市市长对此解释得更为明确：当前黄浦江的状况已经"与现代化的上海不相配"了。

然而，更深层的原因是经过20年的建设，上海的工业迅速发展，城市快速扩张。而今天上海的目标已不仅仅是创造中心城市，更面临产业升级与转型的机会，人们将注意力转移到了追求高质量的生活方面。期待以更好的环境、住房和休闲娱乐设施吸引来自世界的人才，也奖励多年辛勤耕耘的居民。另外，这也是浦东新区全方位发展的结果。市政府官员指出，这一工程能够更好地推动浦东新区的发展，加快城市经济的发展。同时，浦东的成功经验也为此提供了一个发展模式。最后，新的海港建设也要求黄浦江滨江地区转换以前的航行的功能，而将它建成一个集开放空间现代化办公楼、商务楼、住宅区和休闲娱乐设施为一体的区域。

不过，这项工程最直接的动因是为2010年世博会提供所需的场所。世界博览会（EXPO）被称为经济、科学、技术界的奥运会，这是一次大规模、全球性、非商业性的大会，旨在提高和促进世界的发展和交流。自1851年世博会在伦敦举办以来，尚未在发展中国家举行过，而仅仅是在一些欧美及亚洲发达国家，像美国、加拿大、日本和韩国等，所以这次举办对上海来说有着特殊的意义。上海将它视作未来发展的大好机会。2010年上海世博会的主题是"城市，让生活更美好"（Better City, Better Life），目前，大规模的拆迁与新建工作已经陆续展开。

2. 目标

就像纽约和圣保罗分别成为北美和南美的中心一样，上海希望成为亚洲的金融和经济中心，因此必须和东京、香港和新加坡等城市竞争，而世博会的筹建和召开将有助于将投资者从这些竞争城市吸引到上海，同时也能吸引大量的旅游者。从城市形象上说，世博会总能推动城市环境及建筑的发展，而上海选择黄浦江两岸的滨水区作为会址，也期望通过场馆建设的契机，改善当地的旧城区，提升环境品质和生活质量。

就像申博的主题一样，上海广为宣传其作为适合

案 例

图19　上海的申博方案——法国 Architecture Studio

图20　上海2010年世博会规划方案——总平面

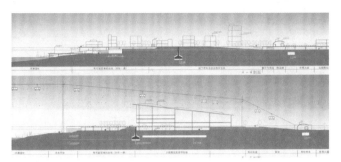

图21　上海2010年世博会规划方案——剖面图

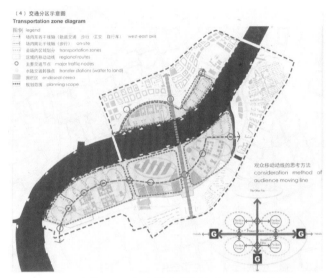

图22　上海2010年世博会规划方案——交通组织

人们居住、生活、工作、生意的场所，并试图将"2010年世博会"作为上海整体发展的重要战略，向世人宣传和推销上海。而实际效果是，作为首个在发展中国家举办如此大规模商务活动的城市，上海已经至少证明自己已经开始获得全球性的关注。如果上海能够在持续数年的筹备中抓住全球化带来的机遇，特别是在展会期间展现出自身在融资、建设、运营、管理、应急、反恐等方面的能力，上海无疑将提升在全球城市网络中的等级，并引领整个中国在全球化的步伐，最终超越南美诸国成为对外来投资者更安全、更有效、收益更高的国家（图19～图22）。

3. 挑战

对于国内经济的发展和全球化进程来说，2010年世博会的到来似乎应该更早一些，国家经济过于倚重出口加工贸易，既忽略了品牌、设计、营销渠道等关系到未来企业竞争力和国家竞争力的要素，又消耗资源及影响了环境的可持续发展。而世博会展现的是一个国家在科技、制造、文化、经济等方面的综合实力，因此，近几年我国在参加国外举办的世博会时展览销售的低品质的加工产品、简单的手工艺品将不能成为上海世博会的主展品。我们既要在场馆建设中展现建筑、环境、景观等领域的成就，更要通过2010年世博会的契机调整国家发展策略，形成独特的经济环境和具有竞争力的产品，向世人展现科技实力和文化魅力。

4. 实施

整个工程的落实需要遵循市场操作和多元化投资的策略。随着筹备活动的开展，市政府只可能提供其中的一部分资金，更多的发展资金将来源于国内外的市场投资。

现在一个注册资金为两亿元的开发企业已经建立起来了。同时，一个由市长担任主任的监督工程发展委员会也已设立。

整个世博会工程将耗资300亿元人民币左右，其中43%来自政府，36%来自各种公司，21%来自银行。而在所有的投资中，58%用于搬迁基地内的企业和居民，42%用于世博村与世博园的建设。

七、结论

上海在短时间实现的爆炸式增长，主要是由于经济改革的政策所引发，许多兴趣不同的行动者妥协的结果。各级政府的角色包括了本地政府和中央政府的协同博弈，以及地方和全球网络的连接，政府也不可避免地受到强大的全球力量影响。这就是全球化的真实写照。虽然这里回顾的案例是基于中国独特的背景，特别是积极主动的金融机制和公私相互作用，但其发展经验同样适用于其他处在转型过程中的社会中的城市。

参考书目

[1] Cartier C. Globalizing South China [M]. Oxford: Blackwell Publishers, 2001.

[2] Castells M. The Rise of the Network Society. Part 1 of: The Information Age: Economy, Society and Culture[M]. Oxford: Blackwell, 2000a.

[3] Castells M. End of Millennium. Part III of: The Information Age: Economy, Society and Culture [M]. Oxford: Blackwell, 2000b.

[4] Chen Y. Shanghai, a Port-city in Search of a New Identity: Transformations in the Bond between City and Port [M]// M. Carmona, eds. Globalization and City Ports. Delft: Delft University Press, 2003.

[5] Chen Y. Public-private Partnership in Urban Area Redevelopment in a Chinese Context: Lessons from Shanghai Pudong Development [M]// C. Feng, M. Yam, W. Lu, D. Drew, Y. Tan, M. Yu, eds. Advancement of Construction Management and Real Estate. Hong Kong: Chinese Research Institute of Construction Management, 2004: 353-364.

[6] Gamble J. Shanghai in Transition: Changing Perspectives and Social Contours of a Chinese Metropolis [M]. London and New York: Routledge Curzon, 2003.

[7] Kirby R. Urbanization in China: Town and Country in a Developing Economy, 1949-2000 AD [M]. New York: Columbia University Press, 1985.

[8] Olds K. Globalization and Urban Change: Capital, Culture, and Pacific Rim Mega-Project [M]. Oxford: Oxford University Press, 2001.

[9] Sassen S. The Global City: New York, London, Tokyo [M]. Princeton: Princeton University Press, 1991.

[10] Wan Z., Yuan E. Toushi Pudong, Sisuo Pudong[M]. Shanghai: Shanghai Renmin Chubanshe, 2001.

[11] Wu F. The Global and Local Dimensions of Place-making: Remaking Shanghai as a World City [J]. Urban Studies, 2000, 37 (8): 1359-1377.

[12] Wu W. City Profile—Shanghai [J]. Cities, 1999, 16 (3): 207-216.

[13] Yeung Y. Introduction [M]// Yeung Y., Yun-wing S., eds. Shanghai: Transformation and Modernization under China's Open Policy. Hong Kong: The Chinese University Press, 1996: 1-23.

[14] Yusuf S., Wu W. P. Pathways to a World City: Shanghai Rising in an Era of Globalisation [J]. Urban Studies, 2002, 39: (7), 1213-1240.

SHANGHAI

Yawei Chen, Hao Hu

Globalization is an imperative force in making Shanghai a world city in the history. Shanghai became a semi-colonial city when China lost a series of wars to western invaders in the second half of the 19th century. These foreign invaders brought advanced technology and foreign investment into Shanghai, and Shanghai was the most highly developed urban area in Asia outside Tokyo (Wu 1999). Shanghai was isolated from the world ever since communist took over in 1949 and became an industrial base of China in the three decades, which followed. It was only after China adopted the open door policy at the end of 1980s, that Shanghai resumed its close contact with the international world. The target of Shanghai's strategic planning is to build the city into a world-class one by 2020.

As the most important economic city in China as well as the income generator for the central government, it was not so sure to what extent Shanghai could be opened to the outside world. Not until the beginning of the 1990s, the highest level of Chinese leadership decided to open up Shanghai, the heart of China's economy, to foreign investors. In this way, China hoped to deliver a message to foreign investors, ensuring that China's opening-up policy would be continued and foreign investors were to be assured of China's economic future and not take their investment away.

In Shanghai urban development increased greatly only after 1990 when the Central Government announced the development of Pudong New Area District. This article will take Shanghai Pudong New Area's development as an example to explain how a Great Urban Project is designed and developed in the context of a transitional society like China.

Large Urban Project: Pudong New Area Development

Pudong Development Initiatives

The development of Pudong was proposed in the early 1980s to meet the challenge of population growth and business expansions. The proposal for Pudong development detailed in the Shanghai Strategic Development Plan published in June 1984 suggests that Pudong should be developed into a new economic development zone that attracts investment with preferential policies for investors such as tax reduction. In the 1986 Master Plan of Shanghai, Lujiazui, the waterfront area in Pudong, was identified as the hinge for Shanghai's future east-west axis and an area of future extension for business district which cannot be sustained by the historic downtown area.

The central government granted Shanghai a fiscal contracting system in 1988 that was previously only exclusively enjoyed by Guangdong and Fujian provinces. This system sets a cap of revenues taken by the central government and allows local governments to retain all surplus revenues. The central government transferred more authority to Shanghai in the early 1990s in taxation and talents residency management policy. In April 1990 the Standing Committee of the People's Congress approved the proposal on Pudong development. On April 18, 1990, Premier Li Peng officially announced this approval (Wang and Xia 2001: 130). The Pudong New Area was granted "sub-municipality" in administrative status, higher than other districts of Shanghai. In China, a higher administrative status meant more independence in public policy and more authority in issuing policies and approving foreign investment projects.

The Implementation of Pudong Development

Since the start of Pudong development, urban planning had been emphasized by local politicians and academics, partly because of lessons learned from South China that chaotic urban development without planning would result in a deteriorated living environment and finally hamper new investment from coming in. Several features are remarkably innovative in the urban planning of Pudong: firstly, the Master plan of Pudong was one of the earliest master plans in China that was established by law and the plan emphasized on designating development strategies rather than rigid numbers or figures allowed by urban planning; secondly, the master planning integrated with sector plan and economic plan in Pudong in order to achieve economic sustainability. Therefore finance-trade zone, tax-free zone, export-processing zone, and high-tech zone were separately designated; thirdly, a so-called "rolling development" strategy was proposed so that development in each development zone should follow the priority order designed by the plan. Since each development zone was very big (with the smallest zone being 28 square kilometres), the government could not afford to invest in all zones at full-scale at the same time. It was found more efficient to invest in targeted areas in each zone. Therefore, the priority area was clarified in each zone so that the government knew where to target public funds first. Furthermore, every year, the government initiates a timetable of priority infrastructure projects it would implement. In this way, investors who come to invest in Pudong would pour their money according to this priority list, even when there is still nothing but just construction sites.

Rules of Market Economy

Shanghai Municipal Government realised that an efficient government that firmly adhered to the rules of market economy could help boost investors' confidence. To ensure a market environment with more transparency, and fair-competition, the local government decided to establish a small, efficient and open system. To ensure the speedy approval of investment projects for the investors, a so-called "one-door service," system was introduced i.e., that all red-tape work could be arranged within Pudong Investment Centre. Another measure was giving away government control to self-regulated professional associations or development corporations. Furthermore, several development companies were set up to take charge of developments in each development zone. Shortly after their companies were established, each developer received a cheque from Shanghai Municipal Government. With this cheque, they could obtain a certain amount of land from the municipal land bureau with a relatively low land price. In this way land was transferred through subsidized purchase to developers. Shanghai Municipal Government, by investing the initiative fund, became a major shareholder of all development companies. The government therefore gained a voice in the development of each zone.

Land Policy

Pudong has carried out a series of reform in land policy, by which the old land allocation system was abolished. After 1993, all organisations and real estate developers needed to obtain land-leasing rights in Shanghai. Land lease fees were calculated based on floor areas that were to be built instead of the old land leasing policy. This change represented the assignment of real market values of land. The Pudong New Area has an area of 520 square kilometres; among which 300 square kilometres is presently still farmland. Through the land bank system, the government first pays off 30% to land owners of the land price as a deposit to acquire the farmer's land and then pays off the rest when actually acquiring the land.

After these development companies acquire the land, they can choose several ways to get investment for further development: joint ventures with private investors; development companies use land as a kind of deposit to get loans from the Bank; development companies could acquire funding by getting listed in the stock market; development companies sell land-leasing rights to private developers.

Financial Mechanism

To attract more investment, Shanghai government publicized a series of preferential policies, such as tax deduction or exemption for investors. Foreign banks were allowed only in Pudong to conduct Chinese currency (Renminbin) business. Chinese companies have been allowed to establish their own foreign trade companies in Pudong since 1990 even though Chinese companies elsewhere were not allowed to conduct import/export business directly with outsiders but had to go through state-appointed trading companies with exclusive foreign trading representative rights. As a result of these preferential policies, from 1990 to 2003, 470 billion Yuan (US$57 billion) worthy of investment were put in fixed assets of Pudong New Area; about 82.3 billion Yuan (US$ 9.4 billion) was channelled into infrastructure construction between 1990 and 2003 (Shanghai Municipal Statistics Bureau 2004). However, only 10% of the investment comes from the central government, the other 90% is raised from both domestic and international investors through an active financial mechanism (Chen 2002).

Public-Private Collaboration

The actors involved in Pudong come from an unprecedented and complicated mix of local, regional, national and international level, ranging from public organisation, public-funded agencies and organizations, NGOs, Banks and most important of all, the private sector. Public and private interaction is active in the whole process of Pudong development (Chen 2003: p 109). For example, more and more foreign urban planners and architects are involved in urban projects in Shanghai via an open bid system, which could hardly be imagined a few decades ago. Other attempt to involve the non-public sector includes the construction of Nanpu Bridge, Lupu Bridge, and some urban renewal projects such as the Orchid Park. Attracting private investments, rather than public-private partnership is emphasized by the local government, contrary to what happened in urban development projects in the US and Western European countries.

Pudong in Transition: From Development to Creativity

The combination of economic growth, urban transformation and environmental sustainability in the whole process of Pudong development has not only resulted in physical transformations but also provides new economic driving engines to sustain the transformation. From 1990 to 2008, the average GDP of Pudong grew by 17.8% annually. Pudong's development created a historical opportunity for Shanghai to transform from a centre of heavy industry into a centre of finance and new hi-technology. Shanghai has become the engine for the growth of the Yangtze Delta. As Shanghai is granted the host city of 2010 World Expo, Shanghai has used this mega-event as a catalyst for urban revitalisation and waterfront redevelopment. The waterfront area on both sides of Huangpu River is planned to become the exhibition, of which Pudong consist an important part of this project. The world expo project, with the theme "Better city, life better" in mind, aims at creating a new urban centrality with exhibition, retail and entertainment function for the city. Creativity, high tech and sustainability became main motto in the planning and design of the project. The effort can be seen in the redevelopment of the old factories and warehouses along the Huangpu River, which usually faced the fate of being demolished. Most of these buildings are now transformed into work shops for artists and exhibitions centres for the world expo and a new base for Shanghai's flourish creative industries.

What is interesting to highlight is that Pudong's current development path is seen as the indicator for Shanghai's future, ranging from economic strategy, urban policy, and financial mechanism to administration and organisation restructure. Many new reforms on regulations in Shanghai followed the success experience of Pudong. Not only its economy benefit but also its urban feature transformed from this giants urban development program.

Planning and Management: Lessons of Pudong

Firstly, the development of Pudong does not follow a traditional public-planning-finance mechanism. The involvement and participation of actors from different background symbolised a new method that emphasizes on interaction. The involvement of private finance, especially the use of FDI along with new sources of financial mechanism, such as stock market and bonds, has contributed greatly to the rapid development of Pudong. The practice of the Pudong project shows a great potential for public-private collaboration.

Secondly, development companies play an important role in the development of Pudong. On the one hand, they represent the interest of the government and are essential implementers of governmental strategies; on the other hand, they operate in a business way to ensure efficiency and effectiveness.

Thirdly, increasing intensive competition among cities requires the local to react faster and more effective to global influence, absorbing new knowledge, international capitals and resources. In the Pudong case, we have observed the efforts the local government made in introducing new governance concepts, city marketing and market-oriented public policy strategies. However, all new ideas and knowledge need to be nurtured or implanted in the local soil. How a city manages to ensure its position and keep its competitiveness in the future will depend on how the city manages to balance the tension of contradiction and interaction between local and global. Both elements have integrally played key role in reforming the administration body at local level in Pudong.

Within a decade, Pudong, a social and economic wasteland in the past, indeed caught up with the other part of the city. It is now regarded as China's Manhattan, symbolizing the vitality and dynamic of Shanghai, the most vibrant city in fast growing China. All forms a sharp contrast to the image of a "forgotten" land in a most charming city in the 19th and most part of the 20th century. Although it is a case study in China's context, the implementation mechanism such as its active financial mechanism and public-private interaction could shed lights to the practice of Large Urban Projects in a transitional society. By no means, this paper suggests that the process of rebuilding Pudong is smooth and painless. However, many strategies implemented in Pudong and the lessons learned in this project can become reference to cities in other transitional society that implement similar Great Urban Projects to enhance their position in the global setting.

圣地亚哥（智利）

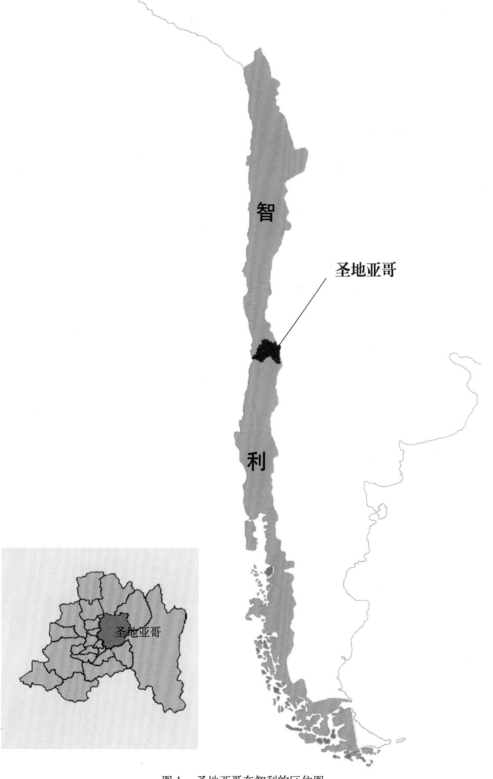

图 1　圣地亚哥在智利的区位图

案 例

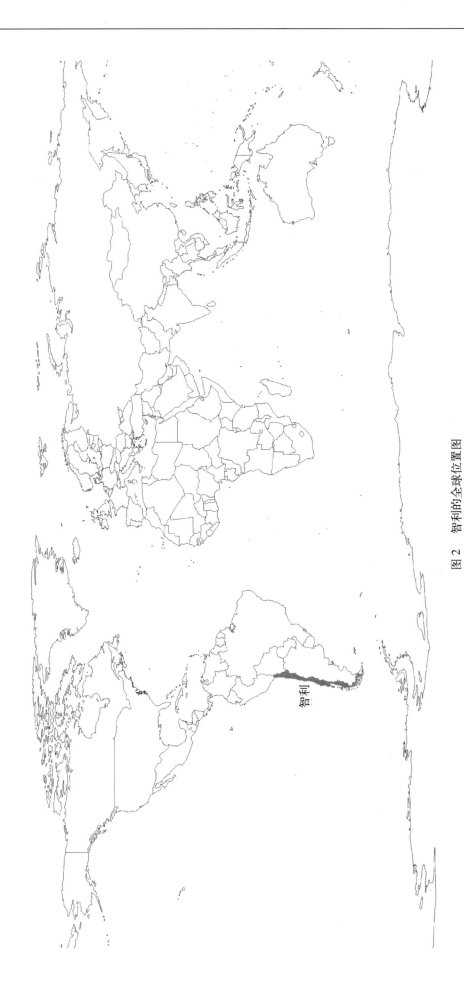

图 2　智利的全球位置图

圣地亚哥

玛丽莎 · 卡莫娜

一、城市数据

（一）圣地亚哥都市圈（SMR）

面积：15600km^2（85%为山峦与河流）。

城区：54个。

人口（2002年）：6061185人。

（二）圣地亚哥市（MAS）

面积：82000hm^2。

城市化区域面积：63000hm^2。

城区数：34个独立区。

人口（2002年）：5413975人。

密度：85.9人/hm^2（城市化区域）。

年增长率：−1.74%（1996年），−1.32%（2006年）。

妇女分娩死亡率：25‰（圣地亚哥富裕区），55‰（南部区域），49‰（全国平均）。

圣地亚哥市占全国GDP的比例：47.6%。

经济活跃人口（EAP）的产业分布（1999年）：

第一产业：4.5%。

第二产业：29.5%。

第三产业：66%。

各阶层人群月收入占全部人群月收入的比例（1996年）：

最低阶层：4.3%。

中下阶层：8.0%。

中等阶层：11.6%。

中上阶层：18.8%。

最高阶层：57.3%。

1. 贫困问题（2000年）

穷困人群：4.7%（2002年），3.2%（2007年）。

贫民人群：12.0%（2002年），9.6%（2007年）。

非贫困人群：83.4%（2002年），88.2%（2007年）。

2. 家庭组成

平均家庭人口数：4.2人/户。

妇女负责的家庭比例：24%。

3. 教育

中学师生比（1999年）：42.5（学生/教师）。

4. 健康

医院床位数（含公立及私立医院，1999年）：3张/千人。

5. 住房

住房套数：1181160套（完整及维护良好）。

住房短缺数（1995年）：319000套（包括已废弃住房）。

社会保障住房数量（1982~2002年）：200000套。

6. 基础设施与交通

机动车数量（1991年）：9辆/百人（在Pudahuel区为2.6辆/百人）。

1977~1991年，乘车出行数量翻了一番。

私人小汽车拥有量每年增长11%。

7. 机动车数

小汽车拥有率：0.75辆/人。

拥有机动车的家庭比例：43.4%。

（三）圣地亚哥市中心

面积：2420hm²。

人口（2000年）：25万人（占整个都市圈的4.4%），每日流动人口超过180万人。

中心区的人口年增长率：-2.0%。

人口密度：103.3 人/hm²。

住房数（1991年）：78530套。

家庭数：96715户（40%为自有住房，20%与其他家庭共同居住）。

人均年收入（1999年）：5105美元。

失业率（1999年）：11.4%。

1. 休闲娱乐

五星级酒店数：15家。

图3 智利中心区域卫星图，包括圣地亚哥及瓦尔帕莱索
（来源：麻省理工学院网站）

图4 被称为Sanhattan的第二中心商务区
（来源：Sepulveda, 2001）

五星级酒店客房数：2766间（占全市全部酒店客房数的38%）。

（来源：El Mercurio, 1998-08-02；Conama, 1999；CADE, 1995；de Mattos, 1999；CIEPLAN/SUBDERE, 1995；CASEN；Mideplan；PGU-MINVU；INE, 2002）

二、城市概况

尽管智利的城市化程度很高（从1950年的63%上升到1998年的88%），人口增长速度趋缓（从1980年的2.5%下降到1995年的1.4%），圣地亚哥市的城市化进程和农村、城市两极分化的现象还是颇为显著。各种政治、经济的手段都难以抵挡城市对农村耕地的侵袭。近20年来，因为市场调节的影响，城市每年扩张达1000hm²。整个圣地亚哥市域由34个独立的城区组成，共有居民约5500万，其中，圣地亚哥市区有23万市民，每天有180万流动人口（图3、图4）。

贸易自由化政策的实施使得圣地亚哥吸引了大量外资。私有化政策又使资本市场在提供退休基金，以及在交通、通信、房地产和公共事业方面提供帮助方面发挥了前所未有的作用，而这些行业对于就业率的增加和GDP的增长都有着重要的意义。据估计，近10年来大约有4000万m²的地皮被用作建设，其中大部分投资在四个城区。该市制订了城市管理方案来降低城市化的速度，并吸收更多的外资。尽管圣地亚哥市的一部分相当富裕，贫困仍是其最敏感的问题，这也是新的城市管理计划所要解决的问题。

圣地亚哥城市规模本没有如此庞大。一直到20世纪50年代，随着低密度的居民区大批建起来才开始飞速发展，尽管那个时候就开始了反城市化的进程。在1950～1980年期间，中心市区大约三分之二的公司都移到了边缘地带。这增加了郊区的人口，促进了那里的就业，也促进了郊区休闲事业和商业的发展。随着汽车的增长，公共交通水平的提高以及道路建设的私有化，反城市化已被提上议程。财富的集中和经济力量的持续增长也促进了反城市化进程。

圣地亚哥住房建设速度下降是所有城市中最高的（每1000个居民每年减少10套住房）。近20年来都是由政府筹建私人承包，且近10年的住房密度相当高（每公顷200户），住房面积也减少到36m^2，价格大约为每平方米150美元，这带来了一系列社会问题。按预计，这种住房普及率的提高以及促进房产二级市场发展的种种新措施将产生类似于20世纪50或60年代的住房迁移，以减少在郊区建设高密度低品质住房用地的需求。

这种城市转型反映了经济、社会和地域的复杂性及其与城市地方化模型的关系。不仅高密度低收入的住房和低密度高收入的住房同时扩张，一些市中心的建设也越来越密集，在南部和中部区域，中低收入居民的住所也愈来愈拥挤（图5）。

圣地亚哥城市的发展体现了其长期的经济增长以及市领导对投资基础设施建设的强大决心。例如，其南北和东西两条长廊促进了该市和南美共同市场的联系。

从长远来看，所有这些工程都将促进社会经济和环境的整合，城市中心以及居民区和工业区之间新型关系的巩固。

三、城市历史

从1541年城市建立到1900年，圣地亚哥的用地大小没多大变化。此后，它的居民区开始扩展，城市中心也变得拥挤和复杂。国家的种种措施，如工业化、交通现代化、地方化、对外部市场的开放、工业转型、反城市化等促使了城市的转型。第一个"社会福利房"的法案在19世纪后半期制定，以控制市区里大型住房的增加。19世纪末，慈善家们发起建造了大批的居住点，形成了现在这种城市居住的模式。第一部社会福利房的法律在1906年通过，旨在提高住所的环境

图5 Providencia 市的快速现代化进程及
土地利用转换 Municipality
（来源：Sepulveda，2001）

条件，并建造了几幢楼房来纪念100周年独立（1910年独立）。根据1925年第308号法案，社会福利高级委员会成立，并建造了43个租用居民区，形成城市花园的格局。1929年，城市发展部成立，以指导民用建筑的总方向，还邀请澳大利亚工程师布鲁内尔来制订城市发展计划。该计划摒弃了美学方面的构想，取而代之以整体化的思路。同时，也提供了处理各种问题的方案，改建市中心，建设新的街道，对城市结构进行分区管理。到了20世纪30年代，非法居住成了一个严重的社会问题，也反映出节约和投资手段的缺乏。1931年通过了补助金法案，该法案作了社会福利房的定义。1933年，圣地亚哥的第一项主要发展计划制订完成。1953年，重建公司和住房银行合并，并制订了一项新的住房计划。于是，社会福利房的建设在土地廉价的地区开始了。各项城市和住房政策也根据不同的方向、发展目标以及国家发展策略制定了下来。根据布鲁内尔的计划，圣地亚哥分别于1960年和1975年制订了两项城市发展计划（图6～图14）。

1979年在军事政府的新自由计划影响下，城市发展开始进入非规范化。20世纪80年代开始，私有化

案 例

图6 圣地亚哥的旧景——Ahumada 街

图7 Alameda de las Delicias

图8 Armas 广场
（来源：Wright，1906）

图10 圣地亚哥东区及中心区一个典型的街区。西班牙街区格局细分后建为社会住宅，称作 Cités
（来源：Boza & Duval）

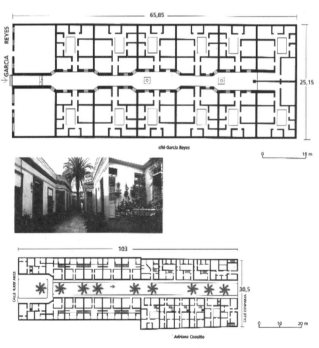

图9 圣地亚哥市传统住宅区平面图及照片
（来源：Boza & Duval）

图11 1876年圣地亚哥规划
（来源：V. Mackenna）

的住房建设蓬勃发展，农业用地被占用。1990年政府重新控制了局面，并形成一套社会福利房和私有化住房一体的体系，以支持低收入居民的住房建设。1994年通过了城市规范计划，确定了城市发展区域和限制区域，以提高土地利用率。

在正确的领导以及开发商们的共同努力下，圣地亚哥的市中心得到了复兴。然而，中心城市的复兴策略并没有用在其他城市的发展上。34个地区

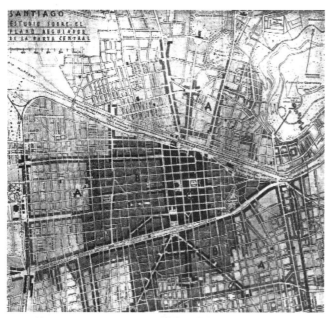

图12　圣地亚哥中心。在1929年由卡尔·布伦纳设计的总体规划
（来源：CA）

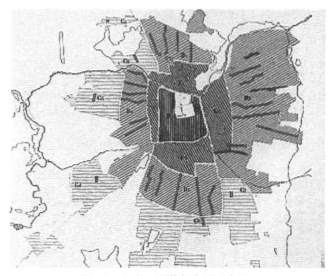

图14　圣地亚哥的都市圈示意图
（来源：CORMU，1973）

图13　圣地亚哥1960年的都市计划
（来源：MOPT，1960）

的48.41%的居民遭受着贫困，其中19个地区中有24000名贫困居民。权利还是掌握在少数人手中，社会不平等现象仍然显著。

四、全球化，机遇和挑战

智利是拉丁美洲国家中第一个打开国门，向外部市场开放的。随着全球化的进程，智利作了相应的财政和经济政策的调整，并认为城市应该向外界开放其产品，商业和金融服务，加入国际竞争。从20世纪70年代起，智利就努力发展经济现代化，采取新的管理方针来加强城市的力量，把权力下放到各城市，提高它们的自主权，并提供财政和管理人员的支援。它鼓励各城市尽可能依靠自身力量，并给社区、本地的企业和房地产商的基础设施和各项服务提供有效帮助。正是这一系列的政策促使了智利经济的繁荣。地方化、非规范化，以及新自由化的经济转型给智利的社会结构带来了巨大变革。近25年来智利的发展经历了两个阶段，在第一阶段中，圣地亚哥的经济活力受到了影响；而在第二阶段，圣地亚哥开始重振雄风。要了解市场开放的社会影响，有必要分析一下雇佣体制的问题。

1967年，第一、第二和第三产业的雇佣比例分别为6.9%、36.9%和56.2%，到1994年则变为4.5%、29.5%和66%。另一个全球化和自由化带来的影响是市区的扩张。10年前，圣地亚哥的用地面积才

$45000hm^2$，而现在有 $60000hm^2$。7% 的经济增长体现在了地方化和高收入居民区的发展上，一半以上的建设集中在 4 个社区，90% 的新建楼房集中在 15 个社区中。也就是说，约有 240 万居民得不到住房。相对比的是，80% 的社会福利房则集中在 11 个贫困地区。对于很多地区来说，社会福利房成了一个负担，因为他们无法满足新生人口的基本需求。

（一）郊区分离化

圣地亚哥居住条件的不平等是显著的。高收入和低收入的地区差距越来越大。1992~1997 年，最贫困的四个地区收入增长了 21%，开支却增长了 37%，而最富裕的四个地区收入却增长达 82%，开支增长了 98%。圣地亚哥市区平均每个居民的开支是最贫困地区的 10 倍，其他地区的 4 倍。在发达地区仅住有 3% 的低收入家庭，而在 11 个贫困地区却没有高社会阶层的居民。圣地亚哥中心的社会条件和其他高收入的地区不同，其中也分布着各个阶层。它是这个国家的行政中心，因此，该地区收入相对更高，而每天的游客人数为居民人数的四倍。尽管已经有国家再分配基金来帮助改变社区间不平等的现状，贫困、失业、犯罪等现象还是愈演愈烈。社会福利房大多集中在贫困地区，无须付财产税，只需一些服务费。这些住房都建在镇上，近 50 年来那里除了福利房几乎没有其他建筑了。低收入的小镇不可能吸引到外界的投资，因而，再分配基金并没有多少可操作性。

（二）边缘地区转型的机遇

我们认为，低收入地区的发展情况和它们的地理位置，交通和通信条件是否能吸引投资，以及有否为商业活动提供援助的计划有关。许多研究表明，圣地亚哥的交通和通信条件和城市职能在近几年中变化迅速。

因地处工业发展区，又因为从内向型工业转为外向型工业和服务业的发展，圣地亚哥的郊区经历了完全不同的命运。很多产生污染的工厂和企业都位于低收入的郊区，和低收入的居民区处于一地。在居民区常常可以发现高污染的化工厂和有毒垃圾的排放。然而，公共交通的发展将给这些地区的交通带来影响，并且正在形成新的中心。一些重要的国家基础设施建设也直接影响了这些郊区小镇的发展。另一方面，新的经济发展圈为之前偏远的小镇带来更多的客流量，使它们成为一些新的城镇中心。地铁线路往南延伸也使交通更为便利。许多途经的小镇都加入到圣地亚哥中心辐射到的圈子里面来。

五、城市战略

到了 20 世纪 90 年代，人们开始注意到 20 世纪 80 年代无规范政策对环境和居住条件的弊端，于是一项新的城市发展计划呼之欲出。

住房和城镇发展部门于 1997 年制定了"圣地亚哥都市规划方案"，阐述了圣地亚哥诸如社会整合、城市特点和功能以及城市发展结构等问题。方案对原先的各社区自行管理的条款作了多项改动，旨在对种种问题通过大范围的、全局的手段采取统一的对策，使得不同地区可以享有共同的利益。这就需要一系列的规范，如下所述。

（一）城市的扩展

城市的扩展需要有规范来确定发展范围、土地使用率、各项目的区域性规划，和国家、地区以及其他部门的联系。通过提高土地使用率来减缓城市化进程，并为社区有秩序地发展提供机会。该方案从整体入手，以人为本，充分利用资源，提高城市效率（图 15）。

（二）公路交通系统

该方案旨在通过区域、城市和社区间的公路交通网发展内外部交通，通过工业和生产活动的地区性规划来不同程度地改善环境条件。地区性规划还要考虑各种发展因素，如城镇就业等，以减少人力资源的浪费，提高交通效率等（图16）。

（三）露天设施

该方案还将通过增加休闲娱乐设施，建设公园、街道、运动场来提高环境质量。

（四）次中心

该方案将在大都市中建设11个小的中心来提高文化凝聚力，增强社区意识等。

（五）生态保护

该方案将对风险地区提供规范性援助，保护那里的生态系统，并根据各地特点合理分配和开发土地资源。

（六）圣地亚哥中心的策略性发展计划

圣地亚哥中心和周围城镇形成一个整体。人们在郊区居住，在市中心工作、学习、购物和娱乐，市中心和郊区通过交通系统紧密相连。这是20世纪90年代政府的一项新政策，尽管反城市化正在进行中，市中心还是得到了新的发展。圣地亚哥中心现为2230hm^2，有23万居民，占整个城市居民的4.39%，而流动人口达180万。

1. 策略性方案

20世纪70年代早期，在华盛顿的支持下，智利

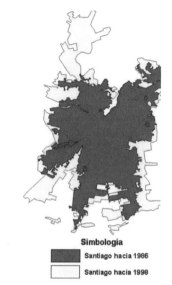

图15　1998年土地利用许可管制

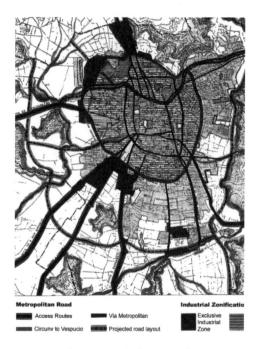

图16　1994年MINVU都市圈规划的道路系统。
第三外环尚未完工
（来源：MINVU，1994）

也主张自由贸易，认为国家的现代化是参与国际竞争的先决条件，因此有必要巩固其作为南美共同市场通往太平洋门户的地位。它认为基础设施现代化是必需的，且可以通过调控政策和自由化实现，这一点在20世纪90年代颇受争议。这种机制和私有化以及和国际金融体系接轨的政策紧密相连。在现代化进程中，大都市得到了优先发展的待遇，交通和人口流动被认

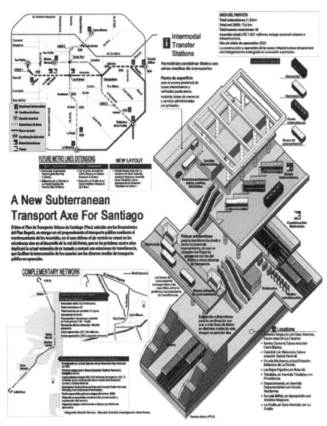

图 17　地铁—公交换乘系统中的车站已成为新的中心，带来商业机会
（来源：El Mercurio，2006）

图 18　中心区的改造程序
（来源：作者）

为是打开国际市场的重要因素。然而，基础设施的私有或公有化在亚洲金融危机时受到了重创。

尽管如此，在金融危机时还是有一系列的工程得到了实施，还有一些在招标。国家的直接干预并不容易，主要因为它不能改变自由化的方针和对市场的透明度的需要。

20 世纪 80 年代早期，水利和电力系统也是私有化的，气候的变化和雨水的不足给水电的输送带来了困难。在智利，91.6% 的人口通了水电。那些贫困地区的小型企业便难逃厄运，因为一到夏季，那里的用水量要增加 10%。

2. 交通发展方案

该方案旨在通过各方面的协调来实现城市和自然环境的和谐发展。如：在城市中建立一个专门的技术机构来控制投资；协调城市发展和环境指数的关系；向南部贫困地区延伸地铁线；为公交系统设立专用道；为所有交通方式制订一张简单的价目表等。该方案的制订是因为不断增加的车辆和交通堵塞现象使得居住环境恶化。而交通问题则是由郊区功能单一化、新兴中心的兴起以及规范化的交通系统的缺乏所导致的。这使得浪费在交通上的时间增加，城市生活质量下降，公共和私有交通系统都受到损失。

另一项措施是加强交通系统的管理（如自动化、路灯的建设等）。通过统一价目表把公交和地铁相结合，私车便可以减少（图 17）。

（七）大都市的复兴

根据政策性方案，城市结构和能源消耗之间必须达到一个最佳的关联。只有这样，经济才能繁荣和多样化，而不削弱商业、金融、管理和文化的各项职能。据称，在大城市，土地使用和交通条件可以通过对城市扩张和次中心发展来达到最佳化。途径之一是减少能源消耗以减轻空气污染，而不是把所有工程都集中在同一地区。该方案包括，加强地铁站的建设，吸引更多外来人员（图 18、图 19）。

图19　中心区航拍图上可以看到其紧凑的形态
（来源：Google Earth）

（八）城市发展规范和小型企业

在圣地亚哥峡谷，农业技术化程度高，土壤肥沃。但由于地理方面的原因，该地区经济发展却很脆弱。因此，土地作为一种稀缺资源必须受到行政保护，比如，控制市区发展，降低城市密度，灵活的地区性规划等。根据2001年6月通过的一项新的法律，中小型企业可以使用免费土地，并且允许其改善基础设施和生产系统。大约3500家企业将从中受益。自20世纪50年代以来，中小型企业的土地使用一直受到限制。土地不能买卖、租用或进行其他改建用途。这项新的法律旨在通过竞争提高企业的生产效率，以促进就业和环境的改善。

（九）中央车站复兴计划

由于铁路交通的重要性降低，居民区的职能不能满足需求，工业区逐渐荒废，原先中央车站以及其附近区域成了圣地亚哥衰落最明显的地区。但是，该地区是城市的东侧入口，在城市地理位置中占有重要地位，因此，1995年该市开始对它的复兴工程。公共铁路公司拥有45hm²土地，除此之外，在其他土地上将建造地铁线路和办公、住房设施。原先的老车站将被改建成商业和休闲中心。该地区将建设成宽阔的林荫道、人行道、绿化带，还有大范围的住宅区和公园。这项工程规模浩大，由住房和城市发展部负责。

六、城市大规模开发项目

（一）"两百周年"计划

该计划是智利最大的城市发展计划，可以和1910年智利庆祝100周年独立时圣地亚哥所完成的城市转型计划相媲美。根据该计划，原先占地245hm²的塞利洛斯国际机场将成为圣地亚哥市南部的新的入口。其他还有一些重要工程在进行当中，为智利的200周年独立纪念作准备。这些工程由国家赞助，许多其他城市也加入到该项目的建设中来。

该计划旨在吸引众多投资商，通过发挥地理优势，为投资者们提供优惠措施，来改善公共环境和居住环境，提高生产能力。为更好地实施这项计划，还需要加强宣传力度，吸收各种创意。

政府表示，其对该计划的整体决策和管理并不意味着房地产业私有化性质的改变，而是通过市区土地的招标，为私有产业提供基础设施和公共服务，来提高其投资热情。这将确保工程有秩序、高质量地进行。

该工程目的在于促进西南部地区的发展，提高城市密度，控制市区向农业区的扩展。通过这项工程，可以减少花在交通上的时间，提高已有服务行业和基础设施建设的效率。洛斯塞利洛斯距离圣地亚哥的历史文化中心仅15min的路程，邻近工商业地区。它也是智利南部海岸发展的要地。

（二）民用和军用机场的重建

在各国，由于城市的发展，特别是贸易发展的需要，原先内陆城市的机场和港口都逐渐遭到了荒废。根据发达国家的经验，那些旧机场可以用来开发房地

图20　航空照片及紧凑形态的中心区
（来源：Google Earth）

图21　旧的塞利洛斯机场航拍图，其周边是圣地亚哥最穷的街区
（来源：MOP，2002）

图22　塞利洛斯旧机场用地作为建城200周年项目基地
（来源：MOP，2002）

产业，由此可以融入到已有的城市发展中。住房和城市发展部还决定借鉴其他国家城市的经验，如柏林、慕尼黑、香港等。机场建设需要大批的土地用来建设基础设施，这将给当地和周围的区域带来巨大的影响（图20～图22）。

塞利洛斯机场建于1930年，它和通往瓦尔帕莱索的公路和铁路以及20世纪50年代形成的海岸线一起使洛斯塞利洛斯和麦普的社区在20世纪50年代由农业化向工业化发展。在工业区背后是居民区，当时那里很多是贫民窟。20世纪60年代采石场的开发也阻碍了居民区的进一步发展。近50年来，那里除了社会福利房的建设就没有建设过其他基础设施和服务行业。工业化的特性和城市的地理位置吸引了大批低收入居民。那里的居民中，约75%属于贫困或极度贫困人口。而中高收入者几乎没有。那里的人均收入为每月186欧元。去年，在麦普社区因为对城市次中心的建设，当地两个社区的生活水平才有了提高。

（三）洛斯塞利洛斯城

洛斯塞利洛斯现在的人口为78696人，人口密度是每公顷49.18人，在工程实施后，该密度预计将提高一倍。尽管当地的风景已经受到了外界的影响，但那里还是有相当的开发潜力。它地处公路交通枢纽，到圣安东尼奥港口，到南部等地都相当方便。然而，地理因素和低收入人口迁入使得当地缺乏内部与内部，内部与外部之间的沟通。

1. 各种方案的征集（设计标准）

2001年圣地亚哥开始征集各种方案，其要求是：首先，为提高城市生活质量，各方面因素，包括环境、风景、建筑和交通等都要考虑在内；其次，在发展市区的同时，还要注意各郊区和社区的发展；再次，必须加强城市南北部和中部的联系，充分利用毗邻中部地区以及国道线的优势，加强社区之间地铁和其他公交设施建设；除此以外，提出的方案要能实现多功能、多样化的发展，充分开发当地的潜力，迎合现代都市生活的多方面需求（图23～图27）。

2. 管理模型：政府统管，私人投资

按照此种管理模式，国家将对该项工程的基础设施建设进行必要的投资，但是私有产业将发挥主要力量。总体性的工程将由一个公私合营的实体来操作以保证工程的整体性。提出的方案要以此模型为前提。

3. 工程的阶段性和发展的灵活性

因为这项工程规模浩大，15年的时间必须分阶段进行，而且到2010年要为200周年的独立纪念日做好准备。工程的计划也必须有一定的灵活度，能够顺应经济发展和市场变化的趋势。

4. 住房工程

约15000套住房将提供给中低收入居民，价格从15725～64285欧元不等。同时，供应的住房应具有多样性。居民区的各项基本设施必须包括各个卫生、教育、体育设施以及商业和文化中心。

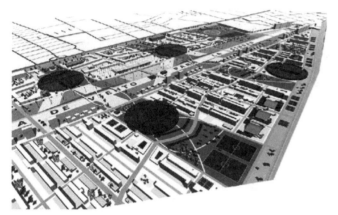

图23　建筑方案竞赛的获奖设计图（2002年）（一）

图24　建筑方案竞赛的获奖设计图（2002年）（二）

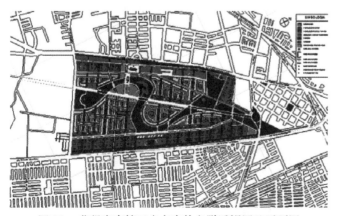

图25　获得竞赛第三名方案的鸟瞰透视图及平面图

图26　获得荣誉奖的方案。所有项目与周边地区
都要有方便的联系（一）
（来源：建城200周年项目办公室）

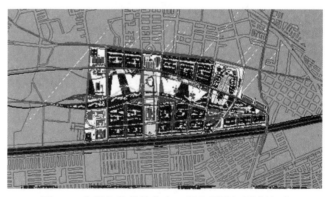

图27　获得荣誉奖的方案。所有项目与周边地区
都要有方便的联系（二）
（来源：建城200周年项目办公室）

5. 公共空间和绿化建设

公共空间的建设是土地使用的一个重要方面。三分之一的土地应该用来建设公共空间，包括公园、公路及人行道等。

6. 总统居所

总统的居住地应该设在合适的位置，既要功能齐全、交通方便，又要考虑安全因素。

7. 洛斯塞利洛斯市中心

该工程还将在市区建设市政府大楼、地方法院、交通局等。

8. 航空博物馆

现在的航空博物馆将保留，其周围地区将预留为博物馆未来发展所用。

9. 其他

在当地还需要建设城市应有的各种服务设施，如医疗机构、体育文化中心以及发展一些无污染的产业来促进就业。

需保留的项目：

（1）塞利洛斯机场的客运大楼；

（2）航空博物馆；

（3）智利空军后勤指挥中心；

（4）高空拍摄的照片中标明的林木带。

七、结论

我们认为，虽然智利与其他拉丁美洲国家存在着经济、政治、社会差异，但全球化所带来的机会使区域内的国家都在经历着结构调整。圣地亚哥因经济的快速现代化和社会的高度两极分化而闻名，其特殊性在于它的城市化形式，以及同时存在的支离破碎的社会结构。智利的第一个悖论是它基于其国家制度和法律体系而形成的高度的集中性，同时却又有着整个南美洲乃至全世界对于市场作用最为放纵的城市管制方式。第二个悖论是由于全球化本身而带来的，即土地管制加强的时候（1990～1997年），城市蔓延问题反而更加严重，而2003年以后土地管制放松的时候，中心区发展反而更加紧凑。目前的管理趋势是尽量对后者的情况加以改变，然而这一目标看来难以实现，因为获得土地最大增值收益的目标已经作为城市发展的准则根深蒂固地渗透到城市体系之中。

看起来圣地亚哥都市圈政府仍然太过弱小而无法根据确立的开发目标来指导新的复杂多样的生活方式，而资源的跨地域流动将持续地存在。

参考书目

[1] Beyer H. Plan Regulador de Santiago. El peso del Subdesarrollo [J]. Estudios Públicos 1997：67.

[2] Castells M. Globalización, Desarrollo y Democracia: Chile en el Contexto Mundial [J]. Fondo de Cultura Económica, 2005.

[3] Echenique M. Algunas Consideraciones Sobre el Desarrollo de las Infraestructuras en Chile [J]. Estudios Públicos, 1996：62.

[4] Delpiano Troncoso, Catalina, Jefa de Proyectos Asociación Portal Bicentenario. Pla de Participación Septiembre 2003-2004[Z]. Corporación Participa. Asociación Portal Bicentenario.

[5] De Mattos, Carlos. Santiago de Chile: Modernización Capitalista y Transformación Metropolitana [M]// En Globalización y Grandes Proyectos Urbanos. Ed. Marisa Carmona. Ediciones Infinito. Buenos Aires, 2005.

[6] Ducci Maria Elena. Santiago: Territorios, Anhelos y Temores. Efectos Sociales y Espaciales de la Expansion Urbana. Reproduced in Eure 2006. Huellas de una Metamorfosis Metropolitana. Santiago Eure 1970-2000 [Z], 2000.

[7] Ministerio de la Vivienda y Urbanismo. Plan Metropolitano de Santiago [M]. Santiago: Publisher Min. De la Vivienda, 1998.

[8] MINVU. Anillo Interior de Santiago. Un Desafio de Gestión Estrategica. Directorio Ejecutivo de Obras del Bicentenario [Z], 2003.

[9] Lawner M. La Vivienda Popular, in Chile hacia el

2000, Tomo II [Z], 1988.
[10] Bossa C., Duval H. Inventario de una Arquitectura Anonima[M]. Santiago: Empresa Editora Lord Cochrane, 1982.
[11] Rodrigues A. Veinte Anos de las Poblaciones de Santiago, Resumen in Proposiciones N.4 [M]. Santiago: Ediciones SUR, 1987.
[12] Hechos Urbanos, Documentacion SUR, 25; 30; 40; 53; 75; 76; 77; 78; 81; 82; 83; 87; 88; 89; 92 [Z].
[13] Parrochia J. Seis Planes Para Santiago. Serie Premio Nacional de Urbanismo N.1. Colegio de Arquitectos [Z]. Editorial Antártica, 1966.
[14] Poduje, Iván (2005-2006) Proyectos y Gestión Urbana. Curso Magíster PUC 2005. In Plataforma Urbana [EB/OL]. www. plataformaurbana. cl.
[15] Poduje Iván. En Defensa del Transantiago. In Edicion Especial On Line [Z]. EMOL. cl. %th July 2007.
[16] Sabatini F. Reforma de Los Mercados de Suelo en Santiago. Chile Efectos Sobre el Precio de la Tierra y la Segregación Residencial. Reproduced In Eure 2006. Huellas de una Metamorfosis Metropolitana. Santiago Eure 1970-2000 [Z], 2002.
[17] Smolka Martim O., Sabatini Francisco. El Debate Sobre la Liberalización del Mercado de Suelo en Chile [Z]. Lincoln Institute of Land Policies. Boston, 2000.
[18] Trivelli P. Power Point Presentation of the Portal de Bicentanario [J]. Ministerio de Vivienda y Urbanismo, 2001.
[19] Valenzuela J. Urban Decay and Local Management Strategies for the Metropolitan Centre: the Experience of the Municipality of Santiago in Latin American Regional Development in an Era of Transition [M]// The Challenges of Decentralization, Privatization and Globalization. Nagoya: United Nations Centre for Regional Development, 1994.
[20] Revista del Colegio de Arquitectos Premio Nacional de Urbanismo.6 Planes para Santiago. CA Santiago [Z], 1996.
[21] Libro Aniversario Dirección de Proyectos Urbanos [M]. Santiago: Ministerio de Vivienda y Urbanismo, 1998.
[22] Rodriguez A., Winchester L.Santiago de Chile, Metropolización, Globalización, Desigualdad: es Posible Governar la Ciudad [Z], 2000.
[23] Bases de Licitación, Diseño de Parques y Espacios Públicos. Proyecto Portal Bicentenario: Construcción de Parques y Espacios Públicos. Gobierno de Chile, MINVU [Z], (Santiago) 2004.
[24] http: //www. portalbicentenario. cl.
[25] Rodriguez A. Las Regiones Metropolitanas del Mercosur y Mexico. Entre la Competitividad y la Complementaridad. Nov-Dic. Buenos Aires [Z], 2000.

SANTIAGO

Marisa Carmona

During the last 20 years Santiago Metropolitan area has suffered substantial changes that keep pace with the persistent economic growth, the country's economic and social stability and the drastic liberalisation of land regulations. The urbanization of 1000 ha per year denotes the primacy of market forces which are simultaneously producing urban sprawl and modernization of the central areas. Greater Santiago is formed by 34 municipalities and about 5.5 millions inhabitants. Santiago central area has a population of 230000 inhabitants and a daily floating population of 1.8 million. New borders of the city express the particularities of Chilean public policies: while high density subsidized housing development are located in the edges of low income municipalities, low density residential developments, are at the edge of the high income municipalities. It is estimated that about 40 million m^2 has been built in the last ten years and most of these investments have been concentrated in only four municipalities. Santiago has experienced a dramatic increase of wealth concentration. Inequality is a serious problem of the city along with environmental predation.

Although central left coalition have ruled for 20 years, economical power continues to be controlled by reduced groups, absolute poverty reduced significantly and social inequalities continue.

City History

Santiago's territorial extension remained almost the same from the time the city was founded in 1541 until 1900. At the end of the 19th century numerous settlements built by philanthropist initiatives took the form of 'cites' and started to shape a particular urban morphology valid until nowadays. The first social housing law was passed in 1906 and several monumental buildings were constructed in 1910, to commemorate 100 years of Independence. In 1929 the Austrian engineer Karl Brunner developed a City Plan which was the first intention of moving away from aesthetic conceptions to planning the city as a whole. In 1933 the Master Plan was approved. Because of the rise of illegal settlements a new instrument by which the concept of social housing was defined (1931) to facilitate subsidy delivery was passed. The city was relatively compact until the 50s, when a double process of social housing construction and counter urbanisation process began. $2/3^{rd}$ of the firms established in the inner-city relocated in the period between 1950 and 1980 (Valenzuela, 1997).

In 1979 with the military dictatorship, the city suffered considerable changes with harsh land deregulation, underpinned by the idea that regulations interfered with the cost and supply of land, and the planned evictions of illegal settlers from the central towards the peripheries areas. A dynamic private housing financial system started and the occupation of agricultural land continued throughout the 80s. The democratic government in 1990 developed an ambitious subsidising system together with the private financial system, supporting the construction of an unique low-income housing program which have built almost the half of the total existing housing stock until the present but in low cost areas.

In 1994 a new Regulator's Plan of Metropolitan Santiago identified the limits of the Urban Metropolitan Area and the Restricted Area. The goal was the intensification of land use. The Santiago city-centre has been regenerated, through ideas of mutually supporting land uses co-ordinated with measures to control mobility. This revitalisation strategy of the inner city has not been replicated in other municipalities. Poverty affects 13% of the population of the Greater Santiago.

Globalisation and Strategic Development Plan

Chile was the first Latin American country to open its boundaries and change from its internally oriented development to the external market. Fiscal and economic adjustments helped cities to become economic units in their own right to international competition. New principles of management efficiency were adopted to strengthen municipality's powers.

The economic growth of 7% per year has been reflected in the double process of spatial deconcentration and concentration of activities. An image modernization and intensification of centrally located municipalities (i.e the 'the Sanhattan') took place where almost 40 million m^2 was built in ten years (Trivelli 2001). More than the half of these were built in 4 districts. More than half of Santiago's Municipalities have not received any building investment. On the other hand, 80% of social housing is concentrated in 11 poor municipalities. For many municipalities social housing is more of a burden than a benefit, since they are unable to tackle the basic needs demanded by the new population. Although Santiago, is different from other Latin American capital city's, present high rates of services coverage, there is increasing inequality between its municipalities.

The Metropolitan Strategy

In 1993, the Metropolitan Regional Government created the main strategy for the 2000-2006 period aimed at improving quality services and enhancing public and private competitiveness performance. The need to increase knowledge management and technologic development seemed to intensify the use of the infrastructure and to lower pressure on the environment. The Ministry of Housing and Town Planning developed the Metropolitan Regulator Plan of Santiago in 1997 to address the problems of social integration, identity, drug addiction and functionality in the Metropolitan region. The idea was to assure a unitary treatment of territorial problems through a wider and integrated vision in which different actors could share the different spaces in pursuit of common interests.

Strategic Projects

The need to improve national and metropolitan infrastructure was one of the most sensible strategies in the 90s. The mechanism identified was a partnership policy involving privatisation and concessions, and the encouragement of national-international investments.

Metropolitan Road System

National improvement of infrastructure was considered important advantages for the opening of international market opportunities.

The metropolitan road system plan also sought social impacts through improvement of environmental conditions and the enhancing of industrial and productive activities. The land zoning system associated with transportation, in addition to existing conditions of nuisance also took into consideration development factors. The main roads system built by partnership is the East West axe (Costanera Norte), the Americo Vespucio ring road and the North-South access. Many roads have directly influenced the conditions of peripheral municipalities.[1] The finishing of the Vespucio ring has significantly improved accessibility to previous peripheral locations. New sub-centres are attracting population and services from all over the city. The extension of the underground lines to further south and the new transport system will change accessibility of the municipalities of the periphery.

Master Plan of Transport: Trans-Santiago

The creation of a unified metropolitan transport system combining underground system and buses; the extension of the underground system towards the southern low-income area of the city; the creation of exclusive lanes for motorised bus transport; the construction of transport nodes for modal exchange; and the creation of a simple fare tariff for all types of transportation are the important strategies of the plan. The plan is based on increasing recognition of the deterioration of environmental standards and living conditions of poor population and the low urban productivity, given the increasing number of travels through private buses and constant traffic jams and air pollution increases.

The Sub-Centres Networks and Revitalization of the City Centre

The Strategic Plan aims to achieve an optimal relationship between urban form and energy consumption. The idea is that it is essential to keep the metropolitan centre economically vital and socially varied, without minimising its commercial, financial, administrative and cultural functions and to avoid the exaggeration of mono-residential functions. One way of attaining the goal of controlling urban form and energy consumption is through poly-nucleated schemes with lower energy consumption indices and therefore lower air pollution rates. The strategy involves the creation or revitalisation of centralities reinforcing of "Metro-stations" and the "Transfer modes stations" through underground corridors and their conversion into alternative attraction of trips from metropolitan areas. The Plan defines a network of 11 metropolitan sub-centres for services fostering cultural integration whilst enhancing land values in deprived areas.

---

[1] The renewed South-access to Santiago that goes through La Pintana, La Granja, San Ramon and the opening of the La Feria, the new Parallel corridor to the Gran Avenida Avenue which goes through, El Bosque, La Cisterna, Pedro Aguirre Cerda and San Miguel are all joining the periphery with Santiago-Centre.

Urban Growth Management and Small Enterprises

In the Santiago valley, agriculture is highly mechanised given the fertility of the land. This area is also ecologically very vulnerable and dangerous for urban functions. In this perspective, land is a scarce resource that is necessary to administer with care, and there is a need to establish control of urban growth and intensification of land uses. This goal does not preclude targeted de-regulations such as the one related to those of not conflicting small manufacturing uses. In July 2001 a new law (No 19, 744), giving free land uses for small and medium enterprises in the whole country was passed. This law permits small and medium entrepreneurs to improve infrastructure and productive systems. It is expected that about 3500 enterprises will take advantage of this law.

Recycling Old Areas and Buildings Project

The area around the Central Station is a degraded area of Santiago. The loss of importance of the national railway system, the shortage of residential functions and the increasing dilapidation of industrial areas are the principal causes of this downgraded situation. Because of the importance of this centrally located area, in 1995 the operation to revitalize the zone was initiated. There are 45 ha owned by the public railway company and the project involves the building of underground lines and of residential and office accommodation on free land. The old station, which was declared a National Monument is today a commercial and leisure centre. It is believed that this project will impact on the surrounding areas between the Mapocho River and the Av. Bernardo O'Higgins.

The targeted image improvement of the area includes the Matucana Ave, as a large boulevard with large pedestrian paths and greens. This is a mega-project and will have 4000 housing units and 10 to 15 ha of parks. The project is being organized and financed by the Ministry of Housing and Urban Development and the railway partnership. The modern Santiago library is build from the re-conversion of an old public warehouse in a run-down part of the city closed to the metro station Quinta Normal.

Santiago's Inner City Ring

This initiative seeks to recover the downgraded parts of the central area, generated mainly by the presence of obsolete trains lines and industrial sub-utilisation. The goals is to compact and intensify the whole city, improving the green network and connectivity, minimizing the need for commuting trips and the reduction of contaminant emissions. Specific goals are spatially reversing socio-segregation through the formation of new mixed neighbourhoods, changing the way the city centre is used. The project has a target population of 124796 inhabitants and a target area of 1.652 hectares. One of the project in the inner city ring is the new Justice Centre.

Large Urban Project: The BI-Centenary Portal

The Bi-centenary Portal has been compared to the transformation done in Santiago for the celebration of the First centenary of the Independence in 1910. The location is the old Airport of Cerrillos (245 hectares). Because of the location and magnitude of the project, it is believed that it will have a huge impact in the city, and it will create a new access (portal) to the city from the South, it will also create a new centrality and will help the upgrading of the most deprived areas of the city. The procedure of formulating the project is to make a public campaign of the proposal, by calling for a competition of ideas to finally tender for the master plan preparation. The government has made it clear that the planning and management involved in the project does not substitute the natural role of the private sector in real estate business. The state will assume a facilitator role in the design and planning phase, will offer investment initiative to the private sector by way of giving greatest business opportunities through the tender of urbanised land supplied with infrastructure and public services. The population of Los Cerrillos municipality is 78696 inhabitants with a density of 50 inh/ha which is expected to be duplicated by this project. The location is connected to a network of important regional highways which are giving the place a good level of accessibility.

Real Estate Developments: Costanera Centre Complex

Real estate development in Las Condes and Providencia will be intensified with the completion of four towers of more than 50 floors which are presently under construction. The Costanera Centre with a height of 300 meters which is developed by CENCOSUD Holdings, represent an investment of US 400 million and a building complex of 600,000m^2 on a site of only 5.5 hectare. This land use intensification represents a great impact in the already saturated urban system which is not being internalized by the private sector. The complex includes two hotels, a shopping mall, multi-cinemas, two super-markets specialised services, offices and apartments. The annual return on investments of the complex amounts to US 150 million, an amount that feeds the boom of tower construction in the area. The tower Portada de Vitacura of Sennerman (Titanium Tower) is also more than 50 floors and is almost finished. Future developments in Parque Arauco include five towers. Chile has already embarked on a new generation of tower buildings inspite of the country's prevailing seismic conditions, in the process lending a new image to the 'Sanhattan' area. What appears true is that the real estate sector is booming across Chile, and even more in this particular part of Santiago.

新加坡（新加坡）

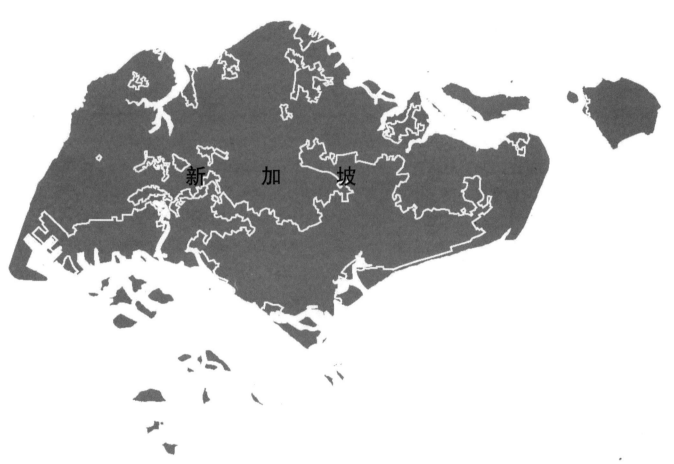

图1　新加坡的区位图

案 例

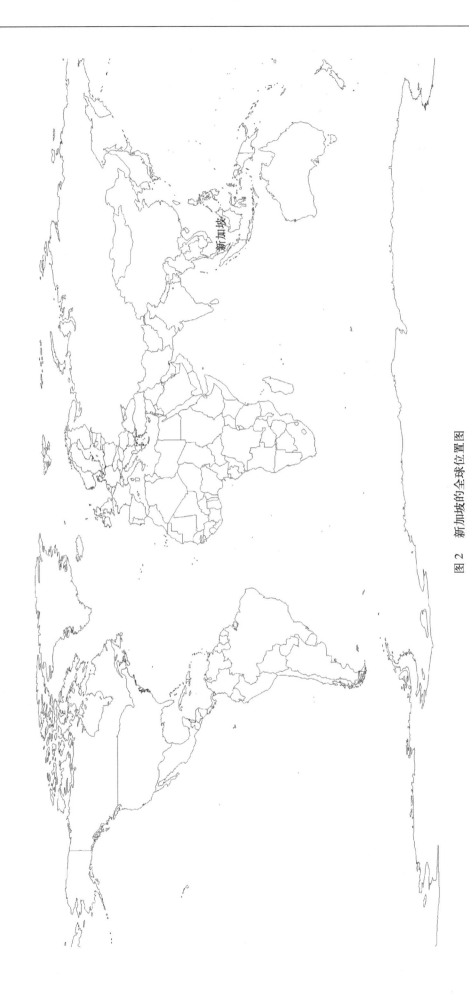

图 2 新加坡的全球位置图

新加坡

胡 昊

一、城市数据

(一) 人口与面积

总人口：448.39 万人。

年增长率：3.3%。

城市人口比例：100%。

新加坡居民：360.8 万人。

年增长率：1.8%。

15 岁以下人口数量：69.81 万人。

15～64 岁人口数量：260.39 万人。

65 岁或以上人口数量：30.65 万人。

出生率（每千人）：10‰。

死亡率（每千人）：4.3‰。

—	总人口	新加坡居民
1990 年	304.71 万	273.59 万
2000 年	401.77 万	326.32 万
2010 年	475.50 万	375.00 万

土地面积：704km^2。

人口密度：6369 人/km^2。

(二) 经济指标

GDP：2099.9 亿新元。

GDP 实际增长率（2006 年）：7.9%。

GDP 实际年增长率（1990～2000 年）：7.7%。

人均 GDP：46832 新元。

外国直接投资（2004 年）：2721.28 亿新元。

新加坡股票市场海外直接投资额（2004 年）：1738.09 亿新元。

储蓄余额：970 亿新元。

进出口总额：8104.833 亿新元。

出口：4315.592 亿新元。

进口：3789.241 亿新元。

年末外汇储备（2006 年）：2105.29 亿新元。

年通货膨胀率：1.0%。

失业率：2.8%。

居民失业率：3.6%。

游客数量（不含经马来西亚陆地口岸抵达）：974.82 万人。

(三) 住房、基础设施及其他社会发展指标（2000 年）

家庭总数：923300 户。

家庭平均人数：3.7 人/户。

平均每月家庭工资收入：4943 新元。

实际家庭收入年增长率（1990～2000 年）：3.1%。

住房拥有形式：

自有住房：93.4%。

租房：6.6%。

居住在公屋中的人口比例：86%。

状况良好的住房套数：917000 套。

居住在住房开发局所建 4 房或以上公寓以及私人住房中的比例：68%。

案 例

住房状况不佳或需要扩建的比例：5.1%。

（四）服务覆盖率

供水：100%。

供电：100%。

排水：100%。

卫生设施：100%。

轨道交通里程（含地铁与轻轨）：93km。

每千人私人小汽车拥有量（2006年）：124辆。

高速公路里程：3122km。

15岁以上居民工作出行方式：

私人汽车：23.7%。

公共汽车：25.0%。

轨道交通：8.6%。

轨道交通与公交：13.9%。

犯罪率（每十万人）：745起。

道路交通事故伤亡率（每十万人）：221人。

（五）教育、健康、娱乐

15岁以上居民识字率（2006年）：95.4%。

会英文的比例：71%。

二、城市概述

新加坡是一个城市国家，是一个由64个小岛环绕的热带岛屿，面积682.7km²，位于马来半岛的最南端，赤道以北1°。主岛的面积约580km²，长42km，宽23km。新加坡的326万人口由多个民族混合组成。1959年英国殖民政府同意新加坡自治。1965年8月9日，它与马来西亚的其余地区分离，变成一个独立主权的国家。

许多年来，新加坡是广为人知的"花园城市"。尽管它靠近闷热的赤道并且城市化程度很高，但由于有计划地种植植物从而使其具有令人愉快的公园般环境。随着水泥森林的增加以及大量道路和高速公路的铺设，新加坡也开辟出来大片绿地作公园和花园，并且沿着道路的伸展种植树木和灌木。新加坡政府不断努力改善环境基础设施并且加强其公共健康控制系统，从而为人们创造了一个世界级的居住环境。现在，它拥有洁净的绿色环境，市民享受着高标准的公共卫生服务。空气和水的污染被控制在世界卫生组织所规定的有利于健康的范围内。因食物、水和传染病所引发的疾病非常少，所有的新加坡人都享受到了现代化的卫生设施。

新加坡不仅地域范围小，而且几乎没有任何自然资源。除了处于国际航空和海运的交汇点这个有利位置外，它没有从大自然得到任何好处（Mahizhnan，1999）。新加坡唯一的资源就是它的国民。然而，新加坡通过利用它宝贵的人力资源而创造了一个"经济奇迹"。仅仅经过一代人的努力，新加坡就从一个居住环境很差、基础设施缺乏的密集城市一跃成为"世界上最繁荣的国家之一"（CIA，2000），人均国民生产总值从1965年的800美元增加到1999年的29610美元，在世界上名列前茅。新加坡已经成为一个为它的居民提供优质生活，并且经济持续增长的优秀热带城市。它的世界级公共基础设施包括位居世界最佳机场之一的樟宜国际机场和世界最忙碌海港之一的新加坡港。在1999年，它的人均国民生产总值位居世界第9位，国民生产总值位居世界第36位（世界银行，2000），经济自由度排世界第2，全球竞争力排世界第1（从1996年开始），世界竞争力连续五年保持第2位，商业市场环境排世界第1，劳动力质量排世界第1，低风险投资环境排世界第2，地理位置对制造商的吸引力排世界第2，并且在最佳商业城市中位居全球第1（Balakrishnan，2000）（图3~图5）。

与其他的城市相比较，新加坡现代化的公屋给人留下了深刻的印象，不断增加的各种式样与规模的高层公屋创造了新加坡独特的城市天际线。在过去的40

图3 新加坡中心商务区的鸟瞰图
（来源：Andrea Peresthu）

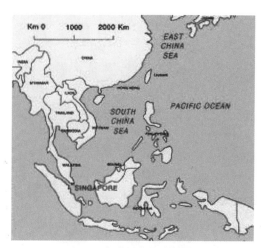

图5 新加坡的战略位置

图4 新加坡民族国家的航空照片

年里，新加坡建成超过 80 万套公共住宅。值得注意的是超过 80% 的新加坡人住在有政府补贴的公共住宅里（现在达到 86%），超过九成的人拥有自己的住房，这使得新加坡的住房自有者的比例在世界上是最高的。这些公屋环境清洁而绿化良好，并且周围有商店、学校、娱乐和社区设备的服务设施。一个由高速公路运输、轻轨交通运输以及公交车组成的公共交通系统，为市民提供了方便快捷地到达岛上不同地区的方式。这些房屋分布在 25 个市镇里，它们中的大部分距离城区 10~15km。公共住宅已经不再被认为仅仅是新加坡人头顶上的一个屋顶，而是"作为经济增长的载体及社会和谐与稳定的基础"（新加坡住房发展署，2000）（图6~图8）。

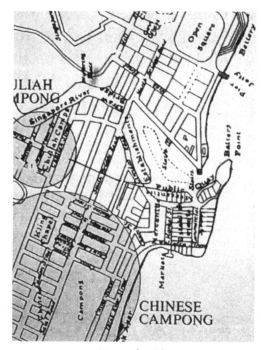

图6 中国城

图7 1830年的新加坡

图8　新加坡传统城市中心

新加坡是一个独特的东西文化融汇的城市，亚洲的传统与现代化自然地共存。新加坡这个多民族社会的人种组成如下：华人占77%，马来人占14%，印度人占7.6%，其他人种占1.4%。许多历史街区表现出不同的种族文化，举例来说，唐人街、小印度区、芽笼士乃、阿拉伯街道等从1989年开始得到特别保护。

三、城市历史

在许多世纪以前，新加坡被称作Temasek，也就是"水城"的意思。依照一些古代传说，在14世纪，一位来自巨港的马来王子在岛上看见一只他认为是雄狮的动物，于是将这座岛屿重新命名为Singa Pura，意思是"狮子的城市"。

现代新加坡是一个英国人，斯坦福·莱佛士爵士建立的。柔佛（在马来半岛）的苏丹于1819年和莱佛士签署了一个条约，将其拥有的新加坡租借给英国建设一个贸易港口。1824年，英国人完全控制了新加坡。新加坡建立后许多移民来此谋生，包括中国、马来群岛和印度的居民。大部分的早期移民背井离乡来到这里，在非常艰苦的环境下工作。他们中很多人从事银行、贸易、农场、造船、警察以及其他许多工作，为19世纪新加坡的经济发展作出了极大的贡献。1867年，新加坡连同马六甲海峡和槟榔岛（现在属马来西亚）一起成为海峡殖民地的一部分，被来自印度的英国人所控制，并成为大多数船只在欧洲和东亚之间航行时停靠的基地，继而成为一个主要的橡胶贸易中心。19世纪后期，新加坡经历了空前的繁荣，它与外界的贸易也获得了飞速发展。

当日本在1942年2月打败英国并占领新加坡的时候，和平与繁荣结束了。在日本占领的3年半时间里新加坡处在一个荒芜的时期，人们无法自由地工作或居住，并导致许多人死亡。直到在1945年战争结束，新加坡重新回到英国人手中。

新加坡人从战争中学到了很多东西，因此不想再被外国人统治。20世纪50年代，新加坡为了获得自治和成立自主政府与英国政府进行了多次谈判，并在1959年获得了主权。人民行动党（PAP）赢得了第一个普选，李光耀成为新加坡国总理。1963年，新加坡加入马来西亚成为它的14个联邦政府之一。然而，联合是短命的，新加坡在1965年8月9日脱离马来西亚成为独立的国家。从此以后，新加坡开始为生存奋斗，并且用自己的双手创造繁荣。随着裕廊工业区的兴盛和其他一些地域的发展，一个大规模的工业化计划开始实施，并为孩子建造了大量学校，政府也为人们建造了许多住房，使他们成为国家的主人。

20世纪70年代前，新加坡已经是一个稳定的国家并实现了全球最快速的经济增长。1981年当时世界上最现代化机场之一的樟宜国际机场建成。1988年，轨道交通系统（MRT）投入使用。1990年，吴作栋成为新加坡国的第二任总理。1968年普选出现的一党制国会变成一种固定模式，在1972～1997年之间的七次选举中人民行动党赢得了大部分或全部的职位。

从一个渔村发展成为全球最发达的地方之一，新加坡的发展主要依靠以下几个因素：绝佳的地理位置、稳定的政府、有责任的领袖和辛勤工作的人民，同样重要的还有这些年来一直在执行的广泛而有远见的规划。

第一个详细的城市规划，就是所谓的杰克逊规划，制定于 1822 年的莱佛士时期。杰克逊规划为各种种族划定了不同的区域，还规划了岛上的商业中心与市民中心，同时还规划了道路网，至今中央商务区还保留着原有的格局。杰克逊规划在其后 8 年里指导了这个城市的发展，但是后来的 100 年间许多建设则像杂草一样混乱地生长在野外（Tan，1999）。结果，在 20 世纪 50 年代有 25% 的人口挤在 1% 的土地之内。殖民政府看到制定一个全面规划以指导新加坡发展的需要，1958 年根据英国新城规划的经验制定了法定计划。这一计划通过区划和密度管理来控制土地使用，并且强调了要满足学校、开放空间、基础设施和社区设备对土地的需求。这个计划摒弃了一般法律对土地使用和发展的控制，从 1960 年 2 月 1 日开始强制实行，并因此成立了一个直属总理办公室的规划机构。新加坡在 1959 年取得自治权，1962 年就邀请联合国城镇发展顾问 Erik Lorange 教授，为城市的复兴做长远规划。先行的两个城市复兴计划就是在 1962 年的 Lorange 规划的指导下完成的（Tan，1999）。1963 年，另一个联合国规划小组由 Otto Koenigsberger 教授率领来到新加坡，帮助他们改善住房问题及防止城市衰败。1963 年的 Koenigsberger 规划提出了起源于荷兰的"环形城市"的方案，建议围绕着中心开放区域，也就是新加坡的中心集水区，建立一系列新镇。

1965 年新加坡脱离马来西亚成为一个独立国家。鉴于新加坡土地资源有限而自然资源匮乏，政府决定发展制造业和服务业以充分利用新加坡的人力资源和它的战略位置。国家及城市的规划于 1967～1971 年间在联合国的帮助下展开，1971 年概念规划正式出台，提出了广阔的长远发展目标和土地使用战略。事实上，在 20 世纪 60 年代之前，1958 年的总体规划就因缺乏弹性而难以对新加坡发展起到指导性作用。

图 9　全球化的影响——市中心的高密度建设

由于概念规划的出现，总体规划的指导作用有所降低，但它仍然具有稳固的法律效力（Foo，2001）。1971 年的概念规划选定的发展模式是在中央集水区周围围绕着一系列卫星城，城镇间由绿色空间和一系列的公园及开放空间作为边界，低密度和中密度私人住宅需要建造在这些城镇周围，此外还为工业区预留了发展的空间。概念规划在 1991 年和 2001 年分别进行了修订（图 9）。

为了将这份整体概念规划变成详细而可执行的形式，总体规划中划定的 55 个规划区分别制订了开发指导规划（DGPs），每个开发指导规划覆盖一个人口约 15 万的规划区域，每个规划区域以后将被分成更小的区域。开发指导规划是一种中期的土地使用规划，它给了建筑师、规划师、发展商以及其他新加坡人一个清晰的概念，就是要建什么、在哪里建、建多高、土地开发强度以及土地的主要功能是什么（Tan，1999）。第一份开发指导规划于 1993 年宣布实施，1998 年修订，并于 2003 年再次修订。全部的 55 个开发指导规划被作为总体规划而刊登在报纸上，作为开发指导规划的补充，还有用于实施的详细规划（DPI）。这些对指定区域的规划是为了使这些区域更与众不同并且改善环境品质。

借着这些规划的指导，许多项目获得实施，以最好地利用新加坡有限的土地和自然资源。举个有趣的例子，我们看到，由于这个长远规划和开垦，新加坡的地形和面积已经发生了变化，由1967年的587km²扩张到今天的682km²，新生的土地几乎有100km²，占原有土地面积的17%，新开垦的土地被用于住宅、工业、机场和新中心的建设上。

四、全球化，机遇和挑战

"每个城市在不同的发展阶段，都有各自要面临的挑战以及解决方案。"（Mah，1999）对新加坡来说，首要的挑战是它狭小的国土面积，对更多私人住房的需求（图10）和社会的多民族化被看做是另外两个主要挑战。

新加坡的主要挑战是土地的缺乏，其国土面积比世界上任何一个中等城市都要小。与其他城市不同的是，它还必须提供作为一个国家必需的空间，例如住宅、工业、娱乐、基础设施、军队训练场地以及机场和港口。更严重的是，对土地的需求会随着经济和人口的增长而进一步增加。居民数预计在15年之内将会达到400万，而在50~60年内会增加到550万，未来对荒地的开垦虽可以将国土面积扩大15%，但由于新加坡的海岸线离邻国的边界不远，它无法过多地填海以取得土地，否则其港口、航道及锚地空间将不敷使用。除此之外，在深度达到15m的水域围垦土地的成本太高（Tan，1999）。另一方面，由于土地的限制，新加坡必须面对许多技术上的制约，例如5个机场旁的建筑需要限高，同时飞机噪声也决定了住宅、学校和医院等无法在靠近机场的地方建设。另外的限制是有20%的国土被用来作为军事用地，缩减这方面的用地会对国家安全造成损害。

还有就是新加坡人比较高的住房需要。一方面，不断增加的移民、更长久的寿命和家庭类型趋向核心家庭都造成了更多的住房需求。事实上，每年需要新建大约2.6万套住房。另一方面，更多富裕的新加坡人对于住房的质量以及个性化有更高的要求（Tan，2000），许多住在公屋的人搬到私人公寓，更多的人也准备如此。然而，就像前面提起过的，新加坡高层高密度的公屋占新加坡住房总量的86%，拥有私人住房的热望是对国家的挑战。如何在这个土地有限的小地方提供更多的私人住房？如何满足不同人的需要和品位？如何应对高收入和低收入家庭的需求差异，在为低收入家庭提供经济适用房的同时又为高收入家庭提供私人住宅（Zhu，et al，2001）？

种族和谐在新加坡也是一个重要问题。新加坡是一个位于伊斯兰环境中的华人占多数的多民族社会，种族和宗教问题在这个地区非常敏感而且非常复杂，以至于这个问题无法在短时间内解决。这种敏感性在许多方面表现出来，比如教育、国防以及对国家统一的威胁。20世纪50年代的种族暴动一直提醒着这个城市更多地关注这些问题。

原则上，新加坡欢迎全球化，因为这对于自然资源匮乏的它来说意味着更多的机会。全球化在新加坡早就不是一个新的过程（Hall，1991），这些可以概括如下：

"新加坡诞生于现代商业资本主义时期，伴随着蒸汽船舶、通信技术以及欧亚间的国际贸易，以及来自东亚、南亚和东南亚的移民而发展。"（Chua，1998）

这一点通过对现代新加坡的发展历程作一个回顾将会看得更清楚。

图10　居住区的大规模建设

19世纪早期，新加坡被视为贸易中转站，不久就"成为东方的贸易中心"（Arun，1999）。这个小岛成功地吸引了很多精明的商人和中间人。在制造业还未发展起来的时候，贸易是它的经济支柱，这一时期被称作贸易经济。20世纪60年代早期殖民关系解除后的政府为了提升它的经济发展和为没有技术和失业的劳动力提供更多的工作，也曾经大力发展过工业。工业化的这一个过程恰逢"新的"国际分工，即跨国公司（MNCs）向发展水平较低的地区寻找市场并且建立新的生产基地，因而得到促进。跨国公司曾被看做是对落后经济地区的一种掠夺和剥削，但是新加坡政府却非常欢迎他们（Arum and Yap，2000）。在20世纪70年代末之前，新加坡坚定地"加入了全球制造圈"（Chiu，et al.，1997）。

在20世纪80年代早期，新加坡已经和中国香港、中国台湾以及韩国在亚洲变成了新经济的四小龙（NIEs）。但是同时，新加坡政府认识到，由于它有限的土地和劳动力，它在劳动密集型和低科技型产业方面的竞争力越来越有限（Chua，1998），于是经济规划者在20世纪80年代开始调整产业结构并强调发展服务业，新加坡开始超越生产中心的地位，进而成为一个国际性的商业中心。劳动密集型、低科技型、低技能型的产业，例如纺织和低端电子等产业逐渐转移到周围邻近的国家。高科技和高附加值的产业，像生物技术、光学技术、信息与通信、先进材料、金融、保险、物流和商业服务业则被鼓励发展。如果将20世纪90年代

图11　市中心的现代化建设

图12　传统居住建筑

图13　战略发展规划（2001年）
（来源：新加坡市政府）

图14　CBD中心区的高密度发展规划
（来源：Andrea Peresthu，2001）

前新加坡的经济称为工业经济的话，那么在20世纪90年代新加坡开始转变成知识经济社会（图11～图14）。

然而，像其他国家一样，新加坡已经发现全球化是一把双刃剑。银行、电信以及制造公司要面对全球环境里越来越激烈的竞争，因为全球化导致了大的公司能够跨国境吞并或者挤垮小的公司。新加坡至今的发展主要是投资驱动的，然而，工资和地价比许多发展中国家要高得多。出口，特别是电子产品的出口是新加坡经济的主要驱动力，但这些产品对全球市场的依赖非常严重，尤其是美国、日本、西欧等国家的市场，这使得它易受全球或区域经济的影响。在1997年亚洲金融危机造成的影响刚刚获得一点点恢复的时候，它又要面对由于美国、欧洲、日本经济发展缓慢所造成的经济不景气。最重要的是，中国的发展对新加坡这种处于发达国家与发展中国家经济之间的一种夹心状态的经济提出了极大挑战。当中国用优质廉价的产品横扫世界的时候，新加坡必须考虑自己如何占有一席之地（Goh，2001）。

在1999年国庆集会时吴作栋总理的演说中，新加坡人可以分成两类，也就是世界主义者和地区主义者。世界主义者拥有全球性的视野，说英语并且有技术，能为国际市场提供产品和服务，并能在世界上任何地方舒适地工作。与之相反，地区主义者仅在国内谋生，其定位和兴趣在国内而不是国外，其技能在新加坡之外没有市场，他们说自己的母语或者新加坡式英语，这些人包括出租车司机、小摊贩、杂货店主、制造工人和承建商。现在面临的难题是让这两个阶层的人相互认同以防止社会分裂，特别是现在全球化和知识经济正在加速这两个阶层的分裂。在世界主义者拥有一个世界级品质的生活时，地区主义者正在为生活而艰苦工作，后者与另一个阶层不断加大的收入差距以及不断增长的生活开支正加深着他们的不满。

新加坡是一个有着大量外国劳工涌入的移民城市，过去10年增加的人口中有一半是外籍劳工、学生以及没有取得永久居住权的外国人，这其中还不包括旅游者和暂时居住者。非国民人口的年增长速度（9.3%）要比国民（新加坡国民以及永久居民）的增长速度（1.8%）高得多。在居民中，增长速度是由永久居民的迅速增长带动起来的（每年增加10%），而新加坡国民的增长仅仅维持在每年1.3%这样一个较低的水平（新加坡统计局，2000）。许多本地人必须和许多比他们受过更好教育且拥有更好技能的外国人来竞争有限的资源和机会，这对维持本地人和外国人的需求平衡带来困难。特别是在经济低迷时期，许多本地人感到外国人抢走了他们的工作机会。因此，许多人呼吁出台一个"新加坡人优先"的政策，并要求政府限制外国人

图15　CBD的现代建筑及传统建筑
（来源：Andrea Peresthu，2001）

图16　跨区域的交通设施
（来源：作者）

的进入。在一方面，政府已经确保新加坡"本地人才优先"，另一方面，政府必须使人们确信新加坡要成为一块"强力磁铁"以吸引外国人才来为新加坡人创造更多的工作机会和财富（Goh，2001）（图15、图16）。

在某种程度上，新加坡面临的这些挑战在其制定的战略规划中都有应对举措。

五、城市战略

尽管国土面积很小、土地资源有限，新加坡仍努力保证一个高品质的生活以及持续增长的经济。新加坡战略计划的主要目标，就是加速新加坡的发展，以使其成为21世纪的一个世界级繁荣都市，一个动态的城市，拥有全球地位的商业中心，一个有特色的城市，有着独特识别性的独一无二的城市，一个充满活力、令人激动和愉快的城市（URA，2001）。

2001年概念规划以550万人口预期以及今后40～50年的增长预期为基础，7份提议在住房、娱乐、商业、基础设施和城市识别等方面提出了关键性的目标（URA，2001），它们包括：

（1）原地改建新住宅；
（2）高层城市生活——每间房子一个风景；
（3）更多的娱乐选择；
（4）为商业创造灵活性；
（5）全球性商业中心；
（6）庞大的铁路网；
（7）充满个性。

这个概念规划的目的是创造一个更适合居住的城市，让新加坡人生活得更舒适，并能提供更多的地段和房型作为选择，居民将享受从城市生活到社区生活，从花园生活到滨水生活的乐趣，他们也可以选择居住在一些小岛上或者在历史建筑里安家，甚至在智能住房里办公，预计未来将建设大约80万套新房。

2001年概念规划的另一目标就是把新加坡建成亚洲顶级的商业和金融中心，包括电子、化学、制药以及生物医学和工程学等高附加值产业将得到优先的发展。主要的金融和服务部门将集中在CBD商圈内以便更好地合作，同时在区域性的中心开发一些商务园区（图17～图19）。

图17 改造后的旧区
（来源：Andrea Peresthu）

图18 中心滨水区的改造（2002年）
（来源：作者）

图19 中心滨水区的改造（2002年）
（来源：Andrea Peresthu）

新加坡也将发展成一个有趣而令人兴奋的城市，提供更多的运动设施供人选择，更多的绿化空间供人们参与到其中的活动，以及在会展中心的更多文化活动。

下面详细介绍正在进行的一些主要的项目。

六、城市大规模开发项目

（一）新的市中心

将在中心区南游艇码头区发展新的市中心，以满足新的扩大的商务活动需要，并将成为靠近现有CBD商圈的一个独立的城中之城。新的市中心将为金融机构及跨国公司总部、五星级酒店和购物中心提供商业用地。建筑物的设计将会充分利用滨水景观，以创造新的独特的天际线（Tan, 1999）。未来将会有更多的住房在这里建造，使居住在城内的人口比例从现在的3%增加到7%（URA, 2001），中心的花园式公寓作为高密度住宅的一种新形式也将提供给市民，目前这是新加坡最重要的战略项目。

（二）智能岛屿

新加坡也许是最先意识到IT业和电信业将带来巨大利益的国家之一，国家计算机委员会（NCB）发表了一份名为《高科技岛国的前景：IT2000报告》的重要文件，报告指出新加坡将利用15年时间发展成为一个高科技岛国，这将会使它成为世界上第一个拥有全国性发达信息基础设施的国家。为实现这一未来的定位，新加坡政府已经开始实施一项大规模的计划以建立必要的基础设施，数十亿新元的投资将注入这个长期的计划。每个家庭至少拥有一条与国家光纤网络相连的线路，国家网络则连接着国际互联网。每一个新家庭必须内置连接到宽带综合服务数字网（B-ISDN）的宽带连接，速度将达到150Mbps。无线网络也将建立，教育部已经计划在2002年为学校里的每两个孩子提供一台电脑，每两个老师拥有一台笔记本电脑。

（三）庞大的铁路网

未来将建设新的环线和放射状线路，放射状线路使人们更直接地到达市中心，环线则使人们在中心区外的两点之间更快捷地联系。捷运系统将延伸到樟宜以及处于东北的一些中心区域，例如榜鹅和盛港。地区间的铁路计划将延伸至南游艇码头区的新城、现有市中区以及城市外的居民点。在城镇内部，地上的轻轨系统也将建立起来。总体上现有的93km轨道交通网未来将增加到500km。

（四）艺术和文化项目

新加坡期望变成一个充满活力的世界级文化都市，政府在2000年3月宣布了文化复兴报告，并且承诺在未来五年内将向艺术领域投入5000万新元。同时，更多的文化和娱乐项目正在计划与实施中。在Marina海湾所建立的Esplanade滨海艺术中心正对新加坡美丽的天际线，占地达6hm^2，是新加坡最大的国际艺术表演中心，总投资达6亿新元。建成的剧场包括1600座音乐厅、2000座剧场、250座独唱音乐厅、220座小剧场、排练间、户外表演空间和购物中心，里面的顶级设备和设施可以与世界上最好的艺术表演中心相媲美，将使新加坡成为国际最前沿的艺术场所，剧场于2002年10月12日对公众开放。

七、结论

在2001年国庆集会的讲话中，总理吴作栋描绘了"新的新加坡"的画面——一个在新的全球竞争环境下、具有社会责任的全球城市。为了实现这个新的新加坡，政府将实现一个新的经济战略并且发展新的社会责任。五大关键要素将推动新加坡的新经济战略：成为全球性的、有进取心的、有更多的创新、进行经济重构、扩大人力资源储备并提升人力素质。新的社会契约——为所有新加坡人接受、在政府和人民之间获得理解——可以保证，虽然经济竞争在加剧，人们的收入差距在加大，但这个国家将会是团结一心的(Goh，2001)。新加坡的未来将成为一个全球性的城市，全球最佳的城市之一，并且拥有使新加坡本国人和全球人才共同生活和工作的最适合居住的城市。

参考书目

[1] Arun M. Smart Cities: the Singapore Case [J]. Cities, 1999, 16 (1): 13-18.

[2] Arun M., Yap M. T. Singapore: the Development of an Intelligent Island and Social Dividends of Information Technology [J]. Urban Studies, 2000, 37 (10): 1749-1756.

[3] Balakrishnan V. Singapore in 1999: A Review [Z], 2000.

[4] Chiu S. W. K., Ho K. C., Lui T. L. City-states in the Global Economy: Industrial Restructuring in Hong Kong and Singapore [M]. Boulder, CO: Westview Press, 1997.

[5] Chua B. H. World Cities, Globalisation and the Spread of Consumerism: a View Form Singapore[J]. Urban Studies, 1998, 35 (5, 6): 981-1000.

[6] CIA. CIA World Factbook[M]. Washington: American Central Intelligence Agency, 2000.

[7] Foo T.S. Planning and Design of Tampines, an Award-winning High-rise, High-density Township in Singapore [J]. Cities, 2001, 18 (1): 33-42.

[8] Goh C.T. Prime Minister Goh Chok Tong's National Day Rally 1999 Speech [N]. Straits Times, 1999-08-23.

[9] Goh C.T. Prime Minister Goh Chok Tong's National Day Rally 2001 Speech [N]. Straits Times, 2001-08-20.

[10] Hall S. The Local and Global: Globalisation and Ethnicity[M]// King, A.D., ed. Culture, Globalisation and the World System. London: Macmillan, 1991: 19-39.

[11] HDB. HDB's Annual Report 1999/2000 [Z]. Singapore: Housing & Development Board, 2000.

[12] Mah B. T. Forward for Home, Work, Play [Z]. Singapore: Urban Redevelopment Authority, 1999.

[13] Singapore Department of Statistics. Singapore Census of Population 2000—A Quick Count [Z]. Singapore: Singapore Department of Statistics, 2000.

[14] Tan G.C. Meeting the Millennial Challenge: Re-inventing Housing, Proceedings of HDB 40th Anniversary International Housing Conference [Z]. Singapore: Times Publishing Group, 2000.

[15] Tan S. Home, Work, Play [Z]. Singapore: Urban Redevelopment Authority, 1999.

[16] URA. Concept Plan 2001: Towards a Thriving World Class City [Z]. Singapore: Urban Redevelopment Authority, 2001.

[17] World Bank. World Development Report 1999/2000: Entering the 21st Century[M]. Geneva: United Nations Publication, 2000.

[18] Zhu X. D., Hu H., Deng L., Huang L. Fifty Years of Public Housing: an International Perspective. Working Paper, Harvard Joint Centre for Housing Studies, Harvard University, Cambridge, Massachusetts [Z], 2001.

[19] http://www.sg/flavour/profile/AReview/indexar.htm.

SINGAPORE

Hao HU

Singapore is a city-state. It is a small tropical island with 64 surrounding islets, which together occupy a total area of 682.7km^2. Singapore is situated at the southern tip of the Malay Peninsula, just 1° north of the equator. The main island is about 580km^2 in size, roughly 42km in length and 23km in width. The population of Singapore is a multi-racial mixture of 3.26 million.

Many centuries ago, Singapore was called Temasek, i.e. "Water Town". According to some old stories, in the 14th century, a Malay prince from Palembang spotted an animal which he thought was a majestic lion on the island. He renamed the island Singa Pura, which means "City of the Lion".

Modern Singapore was founded by an Englishman, Sir Stamford Raffles. The Sultan of Johor (in Malaya), who owned Singapore at that time, signed a treaty in 1819 with Raffles giving the British the rights to set up a trading port. In 1824, The British took over full control of Singapore. After Singapore was founded, many immigrants came here to make a living. They came mainly from China, the Malay Archipelago and India. Many of these early settlers left their families behind and worked hard under uncomfortable conditions. They worked as bankers, traders, plantation workers, boat-builders, policemen and in many other jobs. They contributed much to the growth of Singapore in the 19th century. In 1867, Singapore became part of the Straits Settlements together with Malacca and Penang (now in Malaysia).

The Straits Settlements were controlled by Britain from India. Most ships stopped in Singapore when they sailed between Europe and East Asia. Singapore also became a major trading centre for rubber. In the late 19th century, Singapore experienced unprecedented prosperity and Singapore's trade with the outside world increased greatly.

However, the peace and prosperity ended when Japan defeated the British and conquered Singapore in February 1942. Singapore remained under Japanese occupation for three-and-a-half years, which was a harsh period. People were not free to work or live as they liked and many people died. In 1945, the war finally came to an end and Singapore was returned to the British.

In the 1950s, Singapore held many talks with the British for self-rule and self-government was attained in 1959. The People's Action Party (PAP) won the first general election and Lee Kuan Yew became the Prime Minister of the State of Singapore. In 1963, Singapore joined Malaysia and became one of its 14 Federal States. However, the merger was short-lived. Singapore was separated from the rest of Malaysia to become an independent country on 9 August 1965. Thereafter Singapore's struggle to survive and prosper on its own commenced. A massive industrialization program was launched with the extension of the Jurong industrial estate and the creation of small estates in other areas. Many schools were built for children. The Government also built many houses for the people so that they would all own a part of Singapore.

By the 1970s, Singapore was already a stable country and had one of the fastest growing economies in the world. In 1981, Changi Airport, one of the world's most modern, was inaugurated. In 1988, the Mass Rapid Transit (MRT) System was inaugurated. In 1990, Goh Chok Tong became the second Prime Minister of Singapore. He stepped down in August 2004 to make way for Mr. Lee Hsien Loong. The one-party parliament that emerged from the 1968 general election became the pattern, with the PAP winning most if not all the seats in the nine elections between 1972 and 2006. The PAP is credited with turning Singapore into one of Asia's richest and most modern societies, but condemned by critics for restrictions on dissent.

Singapore is not just small but has almost no natural resources. Apart from its advantageous location at the inter-section of international air and sea routes, it has no other gifts of nature. The only resource Singapore has is its people. However, Singapore created "an economic miracle" by properly exploring and utilising its precious manpower. Through just one generation's efforts, Singapore has leaped successfully from being an overcrowded city with poor living conditions and insufficient infrastructure to "one of the world's most prosperous countries" (CIA, 2000). The GNP per capita increased from US$800 in 1965 to US$ 29610 in 1999, which is one of the highest in the world. Singapore has become an excellent tropical city with a constantly growing economy which provides a good quality of life for its people. It has world-class infrastructure including one of the best airports and one of the busiest seaports in the world. In 1999, it was ranked 9th in GNP per capita, the 36th in GNP (World Bank, 2000); the second freest economy; the first in Global Competitiveness (since 1996); the second in World Competitiveness (retaining its second place for the fifth consecutive year); the first in Business Environment; the first in Quality of the Workforce; the second in Low Risk Investment Environment; the second in Global Location Attractiveness for Manufacturing and the first as the Top Business City (Balakrishnan, 2000). It is interesting to examine modern Singapore's development route. In the early 19th century, Singapore was founded as a trading post and later "became an Emporium of the East within a short period" (Arun, 1999). The island successfully attracted lots of shrewd traders and middlemen. Trading was the mainstay of its economy whilst manufacturing never figured significantly. The economy in this period could be characterised as a trading economy. However, the post-colonial government in early 1960s embarked on an industrialisation programme in order to promote the development of its economy and to provide more jobs for unskilled and unemployed labour. This process of industrialisation was facilitated by the "new" international division of production, where multinational corporations (MNCs) reached out to less developed regions from their home base to seek new markets and to establish new manufacturing bases. The MNCs were ever thought to be predatory and exploitative by many less developed economies, but they were gladly welcomed by the Singapore government (Arum and Yap, 2000). By the end of 1970s, Singapore was firmly "locked into the global manufacturing circuit" (Chiu et al., 1997). By the early 1980s, Singapore had joined Hong Kong, Taiwan and South Korea to become one of four newly industrialising economies (NIEs) in Asia. But at the same time, the government realised that Singapore would be increasingly un-competitive with regard to labour-intensive and low-technology industries due to its limitation in land and workforce (Chua, 1998). Economic planners began to restructure the manufacturing industries and emphasise the need for the development of the service sector in 1980s. Singapore has been moving beyond being a production base to being an international 'total' business centre. Labour-intensive, low-technology and low-skill industries, such as textile and low-end electronics, were gradually shifted to the neighbouring economies. Meanwhile, high-tech and high value added industries, such as bio-technology, opto-technology, information and communications, advanced material, finance, insurance, logistics and business services were encouraged. If the economy of the pre-1990s could be described as an industrial economy, in the 90s Singapore started to reshape it to become a knowledge economy. The transformation of Singapore from a fishing village to one of the most developed places in the world can be attributed to several factors including a superb geographical position, stable government, responsible leaders, hardworking people and far-sighted planning.

Singapore was well known as a "garden city" for many years. Despite its sultry equatorial location and dense urbanisation, it has a pleasant park-like ambience due to its moderating cloak of been greenery. Concurrent with the rise of concrete towers and the laying of a vast network of roads and highways, has the carving out of green spaces for parks and gardens, and the lining of roads with trees and shrubbery. The Singapore government strives constantly to enhance its environmental infrastructure and strengthen its public health control system to build a world-class living environment for the people. To date, it has a very clean and green environment, and its citizens enjoy a high standard of public health. Air and water pollution levels are well within World Health Organisation standards for healthy living. The incidence of food-, water- and vector-borne diseases is very low, and all Singaporeans enjoy modern sanitation.

Compared with other cities, modern public housing in Singapore has achieved impressive results. Nowadays, high-rise public flats of various designs and sizes dominated Singapore's unique skyline. In the past 40 years, Singapore has built over 800000 public housing flats. It is remarkable that over 80 per cent of Singaporeans have lived in subsidised public housing since 1985 (the current proportion is 86%). Nine out of ten own their own homes, which has given Singapore one of the highest home ownership rates in the world. These public estates are clean and green, and well-served by shops, schools, recreation and community facilities. An efficient public transportation system composed of the Mass Rapid Transit (MRT), Light Rail Transit (LRT) and buses provides the residents with convenient access to different parts of the island. These housing estates are scattered in 25 towns, most of which are located at distances of between 10 to 15kms from the city. Public housing is no longer just a roof over Singaporeans' heads, but has also "served as ballast for economic growth and provided the foundation for social harmony and stability" (HDB, 2000).

Singapore is a uniquely multi-cultural city where East meets West, Asian heritage blends with modernity and sophistication co-exists with nature. Singapore's multiracial society is ethnically diverse: Chinese make up 77% of the population, Malay (14%), Indian (7.6%) and others (1.4%). Historic districts representing different racial cultures, for example, Chinatown, Little India, Geylang Serai, Arab Street have been given conservation status since 1989.

However, as a city state, Singapore has faced a set of challenges. The main challenge for Singapore is the scarcity of land. Singapore is no bigger than a medium-sized city elsewhere in the world. However, unlike other cities, it has to provide space for national needs such as housing, industry, recreation, infrastructure, training grounds for armed forces as well as airports and seaports. In particular, demand for land will continue to increase with the growth of the economy and the population. Another challenge is Singaporeans' higher housing expectations. On the one hand, increasing migration, longer lifespan and the trends towards nuclear families means more homes are needed. On the other hand, more affluent Singaporeans have a demand for more, better quality and more varied private estates (Tan, 2000). For example, many public housing dwellers have upgraded to private properties and even more want to do so. Racial harmony is another important issue in Singapore. Singapore is a predominantly Chinese society in a predominantly Muslim archipelago. Race and religion are very sensitive issues within and between the countries of this region and are so complicated that they cannot be resolved overnight. The sensitivity is very easily raised in many fields such as education and national defence, which in turn destroys the unity of the nation.

雅加达(印度尼西亚)

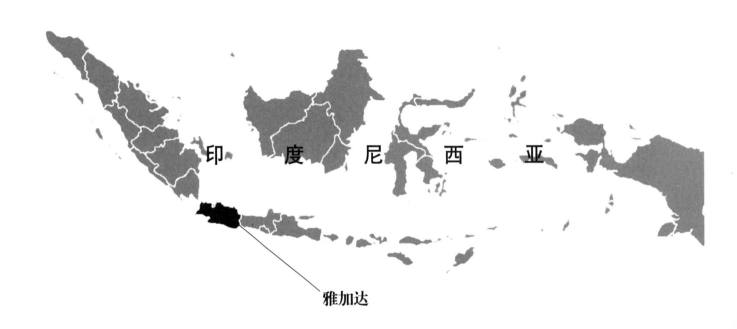

图1　雅加达在印度尼西亚的区位图

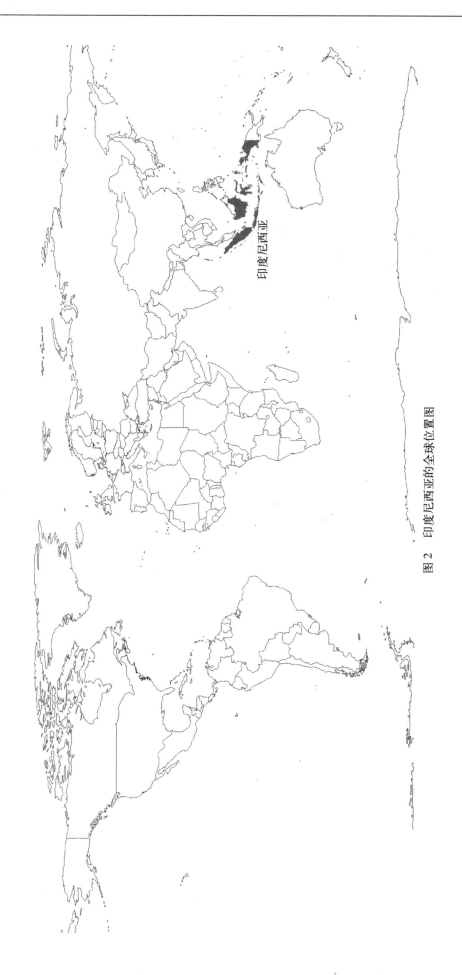

图 2　印度尼西亚的全球位置图

雅加达

安德瑞·派瑞虎 德薇萨俐·特纳斯

一、城市数据

（部分数据根据BPS普查、普华永道报告、印度尼西亚投资促进会报告整理）

面积（市区）：664km²。
人口（市区）：869.96万。
人口（雅加达都市圈）：2409.44万。
市区人口密度：13100人/km²。

（一）经济数据

GDRP：188.03百万印尼盾。
GDRP增长率（2000~2001年）：3.98%。
人均GDP：1935万印尼盾。
外国直接投资（2000年）：1290万美元。
国内直接投资：340万美元。
土地利用：66152hm²。
住宅用地：41331.32hm²。
工业用地：4988.53hm²。
写字楼及仓储用地：6812.75hm²。
公园：1314.23hm²。
其他：11705.17hm²。

（二）房地产数据

写字楼保有量：400万m²。
写字楼使用率：78%。
月租金：10~14美元/m²（在中央商务区，楼龄少于10年）。
零售面积保有量（市区）：15.3万m²。
零售面积保有量（周边都市圈）：34万m²。
使用率：92.90%。
月租金（市区）：35~40美元/m²。
月租金（周边都市圈）：25美元/m²。
公寓数量：29050套。
CBD内：9800套。
周边地段：19250套。
平均使用率：68%。
CBD内：72%。
周边地段：63%。
酒店客房总数（市区）：21500间。
酒店客房总数（周边都市圈）：1000间。
使用率：66%。
平均房价：21~64美元/间。

（三）教育状况

幼儿园：1588间。
小学：3145间。
初级中学：1017间。
高级中学：473间。
科研院所：310家。
大学（公立及私立）：41所。

（四）健康

公立医院：102 家。

产科医院：210 家。

总计病床数：15557 张。

家庭式诊所：537 间。

公众健康中心/诊所：288 间。

（五）人力资源

就业人口：320 万。

待业者：459 万。

学生：140 万。

其他：51 万。

贫穷家庭数：85835 户。

（六）交通及设施

摩托车：160 万辆。

小汽车：150 万辆。

卡车/货车：34 万辆。

公共汽车：25 万辆。

高速公路日流量：131 万辆/日。

高速公路总长度：112.96km。

国道总长度：153.498km。

省道总长度：1325.093km。

市道总长度：4936.888km。

人行道总长度：501.901km。

年起降航班：200882 架次。

年航空客流人数：1333.2294 万人次。

航空货运重量：135234t。

邮局数量：417 间。

电话线数量：168.3087 万条。

供电能力：2319 万 mW。

供水能力：12535L/s。

二、城市概况

在过去的 30 年里，雅加达，作为印尼的首都和一座拥有近 1000 万居民的城市，以它快速发展的城市密集区闻名于东南亚。得益于 1973～1981 年的石油需求高涨和 1986～1996 年新自由主义的发展以及苏哈托执政期间社会和政治的稳定发展，这座巨型城市 1986～1996 年间的年均经济增长超过了 10%，比全国 GDP 增长速度高出 2 个百分点。高速的经济增长带来新的城市化模式，以及人口在城市、区域和国家之间的重新分布。雅加达的人口在 1990～2000 年间也以年均 4.81% 的速度增长，居民数从 1970 年的 450 万增长到 2000 年的 1000 万。

高速的城市发展和城市人口数量的增长导致了城市边界的不断扩张，城市向外蔓延。城市核心区从 1971 年的 340km^2 扩大到 1997 年的 661km^2。特别是 20 世纪 90 年代的房地产膨胀期，城市区域空间高速发展，同时由于银行信用扩大及外来资金涌入房地产领域（如新建 CBD、低密度住宅、工业地产、高速公路等），GDP 中房地产贷款的比例从 1993 年的 6% 增长到 1996 年年末的 16%（64 亿美元）。

然而，1997 年中期，印尼遭受了经济危机，快速发展的雅加达市遭遇多重城市危机。首先是当地货币对美元的汇率下降了 500%，雅加达的 GDP 也从 1996 年的增长 9.26% 下滑到 1998 年的减少 7%。针对危机，政府最初作出的反应是关闭 64 家管理不当的银行，将原来作为长远发展的贷款重新调整为短期银行贷款，金融部门的资金紧缩导致了雅加达几乎所有发展项目的流动资金缺乏。一些特殊领域，如有大量客户贷款的企业和存在大量不可收回货款的企业则遭受重创，甚至破产。在几个月内很多开发商破产，剩下的那些则重新安排或是推迟甚至取消了他们的项目。最终导致的结果是，失业率高速上升，同时也导致了

图3 1937年Batavia商业区的中式建筑

雅加达市内社会和政治的混乱局面。

经济状况的恶化,以及短期内社会和政治的重大变革,最终引发了雅加达历史上最大的城市骚乱。特别是1998年5月13日针对华人社区的那一次掠夺、抢劫,甚至纵火,连续两天的动乱导致了上百名无辜市民死亡,财产损失达2.5亿美元。短时间内,城市经济急速下滑,大量资金外流,同时许多重大投资项目也被取消了(图3)。

两年后,城市的经济缓慢恢复,GDP增长达到4.5%(2000年)。然而,经济恢复还是存在着一系列国际的和地方上的障碍。例如,很多公司由于考虑到投资环境的不安全、政府改革的短视、不断增长的分裂主义倾向以及宗教的武装活动频繁出现,都从印尼撤离(例如索尼电子)。同时,发生在巴厘岛的数次爆炸事件(2002年10月12日、2003年8月5日、2005年10月1日),以及雅加达的爆炸(2004年9月9日、2009年7月17日),也反映了社会稳定正在遭受威胁。

回顾过去,危机都使人们反思当前在新兴发展中国家快速城市化过程中社会与空间发展模式的局限,每一次这样的变化都反映了国际国内在经济、政治、社会、文化和空间关系等方面的转变,以及这座国际性城市中因变化而取得成功或者失败的人群。而社会空间分析的结果却显示雅加达没能很好地处理这些变化带来的影响。在未来的发展中,雅加达需要在城市发展战略中考虑更大范围的可持续性增长,建设更为有效的城市管理体制,采用更为灵活的手段去维持城市的均衡发展,减少国际化发展与地方活力建设的冲突。

三、城市历史

Sunda Kelapa(旧时对雅加达的称呼)在12世纪时是海港城市,它收集和管理来自内地贸易区的商品,并与国际商人进行贸易往来。1602年,VOC(荷兰联合东印度公司)到达这座城市,并将其命名为Batavia。公司利用其强权而很快扩展,并在政治上要求当地政府允许其对香料的贸易市场进行垄断。商业与贸易的发展使雅加达积聚了大量财富,但同时它的内城和外缘产生了分离。城墙内部发生了巨大的空间变化,其中建设了宏大的欧式建筑或中式居住区,而在外缘,主要是当地人居住地及大量种植蔗糖为主的农业耕地,其发展就很缓慢。城市迅速发展导致居住条件的下降,包括缺乏干净的饮用水、霍乱风行,同时环境被污染。城市环境的恶化促使VOC将雅加达向南扩张。

1800年,荷兰政府取代了VOC在雅加达的统治,城市的经济发展呈下降趋势,并出现金融危机。1830~1870年,经济改革政策取消了垄断,加强了糖、咖啡、橡胶、烟草等农耕业的发展,并采取了自由开放的政策,使殖民主义经济向资本主义转化。逐渐地,本地的经济开始发展,一部分本地资金开始投资于工业、金融业及农业,城市空间获得快速发展(南—东轴线),新建了欧洲公司办公楼、工厂、居住区,此外还新建了一系列基础设施,如深水港(Tanjong Priok,1877)、铁路。

1930~1934年世界性的经济萧条导致了国民收入的减少,随之而来的就是大型公共设施建设和城市维护资金的减少,这直接导致了城市空间质量的下降,特别是环境极度恶化的本地贫民密集居住区,被称为Kampong。考虑到环境的恶化会导致一系列社会和政治矛盾,市政府一直尝试改善本地人居住区的环境。随着经济缓慢地恢复,非荷兰的资金大量投向制造业(汽车业、纺织业)和采矿业(石油、锡业)。大中型制造业所积累的大量财富促进了基础设施和公共设施的发展，例如重新设计的市民中心（Koningsplein）强化了市中心的功能，建设了卫星城镇（Gondangdia 和 Menteng），新建及拓宽了道路以满足工业增长的需要。经济增长与城市空间的迅速变化因为 1942～1945 年日本占领而停止，又由于 1940～1945 年的独立战争而滞后。

印尼独立后，第一任领导人苏加诺将 Batavia 改名为雅加达，建国初期的一个重要特点是自由资本主义转向国家资本主义，国家接管了荷兰公司在一些重要战略部门（金融、基础设施、采矿、贸易）的资金。这些部门国有资金的大量产出被用于偿还国外贷款，以及用于建设大量改善民生的项目（公共工程、交通、医疗、教育）。随着新项目的实施及苏加诺对国家发展的高瞻远瞩，雅加达成为一个新的时期国家奋斗、发展、繁荣的重要标志。一系列纪念性、公共性的建筑（包括体育及会议中心、高层写字楼）落成，以及基础设施（雅加达—茂物的高速公路、南北环路）投入使用，这一切不仅提供了大量的就业机会，同时激励了人民的自豪感，建设了尺度巨大的空间标志物。

1965 年后产生了政治和经济危机，在 1967 年的临时立法机构选举中苏哈托取代苏加诺成为第二任总统。苏哈托执政期间，社会和政治生活的稳定成为主要任务，经济方面则是以恢复为主。随着时间的发展，投资环境逐渐转好，国际组织也恢复对其进行援助，提供软性贷款用于公共基础设施建设，例如 1974 年的贫民居住区改善项目。与此同时，1967 年雅加达提出了都市圈总体规划（1965～1985 年），为区域经济发展战略、行政范围扩张、基础设施建设、住宅发展计划、工业发展以及在建成区外围建设绿带等一系列问题提出空间发展的指导方针。在总体规划的指导下，得益于 20 世纪 70 年代大量海外资金流入、石油价格上涨（1973～1981 年），城市建成区迅速膨胀，超出了原来的边缘，沿着东西轴线发展（图 4）。

这一规划本身是逐步调整与发展的，城市空间的

图 4　20 世纪 60 年代的雅加达
（来源：M. E. de Vletter）

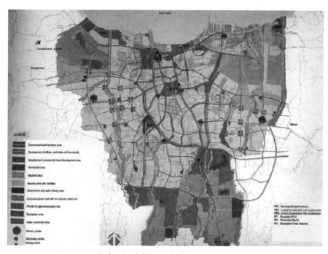

图 5　雅加达总体规划（1985～2005 年）
（来源：市政府）

发展往往会突破已有的规划。由于存在一些不利的空间影响，更为详细的规划于 1974 年编制完成，称作雅加达都市圈发展规划（JMDP）。它指出，提高都市的 GDP 水平需要通过鼓励经济在空间上的分散发展，因此规划提出两个战略：第一，通过大规模的功能区建设形成新的中心；第二，通过建立和扶持位于小城镇的小型企业，推动农业、服务业和就业发展，以迅速提升公共服务、提高居民收入、改善雅加达市内广大的半城市化区域的生活水平。随后，雅加达总体规划（1985～2005 年）对分散式发展的战略进行了详细阐述（图 5）。

然而，总体规划（1985～2005年）中提出的分散发展战略也引发了空间的过度拥挤，而非可持续的增长。特别是在新自由主义政策期间（1986～1997年），大量海外资金的涌入并投资房地产及基础设施领域，导致雅加达的发展超出其空间发展纲要的范围。城市核心向外蔓延，土地投机泛滥，城市通勤人流过密，导致了空气污染、交通堵塞、耕地减少，大量集水区被城市扩张、工厂、娱乐中心、高尔夫球场或赛车场及高速公路所占用。为了实现城市增长，市政府并未关注由此带来的社会和环境的负面影响。结果，20世纪90年代产生诸如城市收入不平等、社会分隔、文化识别性的缺失、环境恶化等问题，并在1997年的经济和政治危机后更加恶化，雅加达1998年发生了历史上最大规模的城市骚乱（图6）。

图6　城市中心的高强度开发活动
（来源：市政府，2005）

图7　亚洲金融风暴对城市的房地产项目产生重大影响
（来源：亚洲城市的转变，2005）

四、全球化，机遇和挑战

由于20世纪80年代中期经济衰退和石油价格下跌，雅加达的GDP年均增速从1967～1972年的11%下降到1982～1985年的3%，政府采用了新的策略以开放市场和鼓励出口，而不仅仅是依赖油气收入。同时，跨国贸易者也希望通过WTO（世贸组织）、GATT（关税及贸易协议组织）、AFTA（东盟自有贸易区）以及APEC（亚太经合组织），来获得更为自由的贸易和市场机会。因此，印尼作为成员国需要进一步开放市场来兑现它对国际贸易的承诺，在一些关键性领域放松国家管制，特别是金融（银行、证券市场）、工业（汽车业）和基础设施（港口、机场、通信、道路）领域。几年内，大量的外国直接投资投入这些领域，推动了GDP年均增长率超过9%，相对照的是1982～1985年的年增长率只有3.1%（图7）。

外国投资在20世纪90年代初期增加明显。例如，1989～1990年两年时间里外资在181个项目中达到13.81亿美元，而1979～1983年的四年时间里，大约只有38个项目共计9.633亿美元。换言之，经济改革后，外国流入的资金数量迅速增加。此外，大多数外国投资都来自于环太平洋国家和地区，特别是日本和香港。1967～1990年间，日本对雅加达的投资占39.8%，而香港占12.8%，其余海外资金来自于欧洲（20%）、美洲（8.5%）、澳大利亚和新西兰（2.4%）、非洲（0.6%）和其他（5.9%）。因此，1990～1997年间GDP的年增长率超过15.2%（图8）。

经济不断发展导致了城市空间的深刻变化，雅加达经历了城市中心与外围极大的变化。在城市中心，由于外国直接投资涌入房地产领域（新写字楼、商务摩天楼、豪华饭店、高层公寓），大规模项目不断涌现。1982～1992年间建成大约600栋新楼，每栋投资约1000～5000万美元，外国投资占了约100亿美元。同时，城市中心区在空间上及经济上经历了转型，

图8　放松管制的政策加速了发展
（来源：亚洲城市的转变，2005）

图9　高密度的住宅开发项目
（来源：Andrea Peresthu，2004）

其余的城市增长方式包括城市边缘蔓延、形成新市镇、建造大量低密度高收入住宅。大量土地被批给私人开发者（121629hm²），而仅有11%左右（约13600hm²）已经开发，超过100000hm²尚空置。此外，由于外资推动的工业需要，已批出9000～12000hm²工业区用地，其中69%左右已经开发（图9）。

很快雅加达就成为该地区新兴的全球城市，国内外许多公司，以及许多跨国公司的分公司都来此经营。因此，城市需要最好的服务、基础设施、通信及技术，这使城市经济变得强大并且使雅加达吸引了大量资金和人才资源。

经济发展以大量的外国资金流入当地许多部门和显著的空间发展为标志，然而，当东南亚经济危机袭击印度尼西亚时，这一切突然停止了。1997年亚洲经济的衰退，导致印度尼西亚经济由于本国货币的显著贬值而崩溃。由于房地产领域使用了大量的外国短期贷款用于其长期而有风险的项目，本国货币的贬值造成极大的损失。雅加达的经济活动紧缩立即引起许多项目推迟或终止，大多数住房开发商倒闭或者撤销他们的开发工程，仅房地产领域就有约16亿美元的贷款无法偿还（1998～2001年）。同时，国内利率上升导致很多年轻夫妇无法按月偿还其住房贷款，他们中大多数决定放弃这些房子。很多公司或是破产或是业绩下滑，已经导致写字楼需求减少过半，很多人失业，并形成城市新的贫困阶层。

因此，那时的雅加达从财富之城变为危机之城，城市内遍布空置的写字楼、大量荒废的土地以及推迟的房地产项目，即使在CBD里也有一大批烂尾楼。无数无家可归的人在桥下、高架路下安家，或者流浪度日。没人知道哪一天经济能开始恢复，哪一天他们能够离开这里找到他们自己的家。

五、城市战略

（一）雅加达总体规划（1985～2005年）

雅加达总体规划（1985～2005年）对前一轮的规划进行了长时间的修正。为了缓解中心区的过度拥挤问题，规划强调了东—西走廊的空间发展模式以疏解经济与空间发展活动，形成新的发展中心。该规划里的主要政策包括：①将2005年年末的城市居住人口控制在1200万以内，并使人口增长与就业配给相协调；②提高城市土地利用强度，每年平均安置26万居民；③为低收入群体提供经济援助和社会福利；④确保低收入者能够有方便和高效的住宅，以降低他们的交通费用和获得较好的城市设施服务；

⑤限制水资源的使用，并且限制城市北部水资源保护区的新开发活动，保护城市南部的蓄水区；⑥协调城市中心与周边区域的整体发展（Bogor-Tangerang 和 Bekasi）。

（二）城市基础设施发展整合计划（IUIDP）

作为国家城市发展战略（NUDS）的组成部分，城市基础设施发展整合计划（IUIDP）从 1990 年年初开始启动。IUIDP 内涉及的基础设施主要包括供水、排水、固体废弃物处理、防洪、地下管道系统、破旧居住区基础设施改进（KIPs）、市场基础设施改进计划（MIIPs）和城市道路建设。IUIDP 按规划、策划及实施的需要分别予以安排，而地方政府则可以自行制定基础设施发展的五年中期规划（PJMs），其中必须包括：技术安排（规划、可行性分析、工程安排）、融资安排（资金需要、融资来源、成本回收），以及实施这些计划的机制需要。然后将这些规划转化为具体的行动计划和预算方案，协调本地、省和中央财政的投资，让公众和私营企业共同参与。最后，城市可以从开发银行或直接通过中央政府获得贷款。

六、城市大规模开发项目

（一）雅加达的外环路项目（JORR）

为了连接雅加达市中心和它的周边新镇，同时减轻通向城市中心部分道路的拥挤程度，政府实施了雅加达外环路（JORR）工程（图 10）。JORR 最初由国有公司 PT Jasa Marga 和 3 家合资企业以 BOT（建造-经营-移交）的方式共同建设。工程的总长度为 210km，并且分为 3 个部分：西部（从 Pondok Pinang 到 Kebun Jeruk）；东部（从 Taman 的小交叉口到 Pondok Pinang）；东北部（从 Cikunir 到 Cilincing）。由于金融危机的影响，工程只完成了大约 30% 就在 1998 年终止。工程最近由印度尼西亚银行重组机构（IBRA）接管。为了重新启动项目，该银行重组机构与 PT Jasa Marga 建立了一家合资企业以寻找新的投资者。最近，重组机构确认了一家以 DRB-HiCom 牵头的马来西亚财团来接管这项延迟的工程，新的投资预算达 5 万亿印尼盾（约合 5 亿美元）。

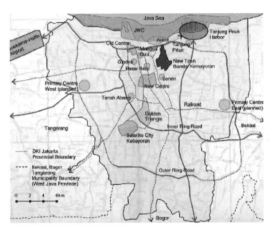

图 10　雅加达外环路项目规划
（来源：BPPK）

图 11　北雅加达湾的填海计划
（来源：www.bappedajakarta.go.id）

（二）北雅加达湾的填海计划

基于 1995 年的 52 号总统令，雅加达开始拟订滨水地区建设计划。雅加达滨水地区管理董事会（JWMB）代表雅加达市政府制订了总投资达 20 亿美元的填海计划，涉及雅加达北部海湾海岸线长 32km、面积 5200hm^2（其中海域 2500hm^2，沿海土地 2700hm^2）的区域（图 11）。该区域预计将容纳 150 万居民，并分为 3 个规划区：中心部分将建成一个新的中央商务区；西部是居民区；东部是海港和出口区。根据 JWMB 的计划，项目预计 30 年完成，并且极大地改善当地的经济和环境状况。然而，项目由于 1997 年的经济危机而推迟。

（三）Kemayoran 数字城市

硅谷在全球的成功已经引发东南亚各国的同样意愿，雅加达也不例外（即使是危机期间）。2000 年雅加达制订了占地 44hm² 的数字城市计划，但还没有实施。数字城市位于以前的 Kemayoran 机场，并由印尼与日本的合资企业经营（印方占 57.5%，日方占 42.5%）。由于这一项目旨在成为国家电子商务、软件、通信、多媒体的中心，区内功能将侧重轻工业、教育及培训、酒店、休闲、娱乐和居住等用途，总投资约 20 亿美元。

（四）社会安全网络计划（SSNP）

这一社会安全网络策略在 1998 年提出，原为保护社会经济危机中受到影响最大的弱势群体。世界银行通过了金额 6 亿美元的短期救援计划，以支持社区基础重建计划，它包括一个重要的本地基础设施维护项目。由于雅加达都市贫穷，世界银行和地方政府官方都努力把此项目与以前的贫民居住区改进计划相结合。其实，很多贫民区早在苏哈托执政期间就改造或者夷为平地了，新近的审计报告还提到诸如专款被挪用、改造目标不当甚至虚构改造项目的现象。

（五）公共汽车专用道

公共交通仍然是雅加达的主要问题之一，虽然道路容量有限，私人小汽车占用 86% 的道路却只解决了 43% 的交通量，而余下 57% 的人依靠公共交通，却只使用 2.5% 的道路。为解决这一交通拥挤问题及提高道路使用效率，政府开始考虑公交专用道作为公共交通规划的一部分。2002 年，地方政府通过了一项计划，旨在西南密集区修建一条长 12km，拥有 93 个车站、60 座桥、地道及人行横道线的公交专用道，耗资 920 万美元。这一项目首先由 50～60 辆公交车以 5min 的间隔服务，预计每小时能运送乘客 10000～27000 人。然而，由于公众质疑这一公交专用道对于缓解交通拥挤的实际作用，同时由于融资困难，市政府推迟了这一项目的实施。（《雅加达时报》，2002 年 10 月 10 日）

（六）大运量快速交通（MRT）

在 20 世纪 80 年代末之前，由于汽车市场的放开，雅加达的私人车辆数量急剧增加，导致交通堵塞。对此，地方政府委托了 3 位外国合作者，即 JICA（日本国际协力事业团）、世界银行和 GTZ（德国技术合作公司），各自完成了他们对雅加达大运量快速交通发展的规划（1988～1991 年）。当这些可行性研究提交后，政府发现很难对每个方案进行评价和取舍。1993 年 6 月，政府将这 3 个规划整合为一个规划，即雅加达城市大运量快速交通发展规划。

根据该规划，大运量快速交通（MRT）工程计划长度为 19km，耗资约 13 万亿印尼盾（合 13 亿美元）。工程选择重轨系统，从雅加达旧城（城市北侧）到 Fatmawati 大街（城市南侧），共设 13 个车站，其中 7 个位于地下。工程设计容量为每小时可运输乘客超过 45000 人，并使出行 20～30km 的时间少于 1h。同时，工程预计能为超过 60000 名建筑工人提供就业机会。

1995 年，印尼中央政府、雅加达市政府以及私人财团（印尼—日本—欧洲资金组成）就该快速交通工程的建设签署备忘录，日本政府承诺向该工程提供总额 15 亿美元、年息 7.5% 的长期贷款（为期 40 年），每年需偿还金额为 3500 万美元。遗憾的是，该项目在1997年印尼遭受金融危机后也推迟了，因为世界银行对印尼政府的救援方案中要求在危机期间终止大规模项目，尤其是使用外资贷款的项目。1999年年底前，印尼政府要求重启该工程，但世界银行依然不同意。在这种情况下，雅加达市政府提出了另外的融资方案，即市政府发行总数为项目投资22%的债券，而余下的78%则由中央政府提供。

考虑到项目的规模，很多研究认为MRT项目的造价及运用费用高昂，因而在印尼并不可行。有些城市的经验则表明，通过结合地下及地面的方式可以显著地降低造价，提高项目的经济可行性。例如台北（1998年）、布里斯班（1998年）、汉堡（1999年）、吉隆坡（1999年）、的黎波里（2000年）都选择了地面系统，以实现比较合理的造价（每公里造价40万～100万美元）。其他一些大都市如马尼拉（1999年）、曼谷（1999年）、伦敦（1999年）、波多黎各（1998年）都选择了高架方式（每公里造价约480万美元），虽然比地面系统昂贵，但仍然比地下的系统便宜得多（每公里造价约800万美元）。

雅加达的MRT工程与区域内其他选择地面与地下结合的城市MRT项目相比造价也要高一些。例如上海的地铁（1999年）总长17km，全线为地下方式，共有车辆144节（每节140万美元），总造价仅2.926亿美元，而雅加达仅半程为地下方式，其造价却为上海的5倍多。香港的西线地铁总长度约30.5km，共有250节车厢，38%为地下线路，其重轨线长为上海的2倍，总投资仅6.2亿美元，不到雅加达的一半。如果我们假设理想的配置标准是每公里轨道线上9～10节车厢（共需155～160节车厢）、每节车厢造价200万美元的话，雅加达需要3.2亿美元（占总投资的21%）用于车厢，而其余的11亿美元用于建造和维护费用，折合6200万美元/km。如果确实如此的话，项目的造价将成为民众诟病的最大问题。

而且，维护成本对于MRT系统来说也是重要的问题，出于政治上的原因，主要是中低收入阶层每天使用MRT，因此收取的车票费不能太高。例如日本东京帝都高速度交通运营团（TRTA）有着超过60年营运地铁的历史，共管理东京8条地铁线路，总长度171km，拥有2431节车厢，至今还无法依靠车票收入来平衡成本，仍然需要地方及国家政府提供补贴，从1961年至今，补贴总额累计高达320亿美元。

与JORR项目相比，MRT的投资额要高2倍甚至更多，但是与MRT不同的是，JORR对私人小汽车交通更有利，而非公共交通。MRT每年能帮助雅加达减少约9亿美元的环境污染和交通拥挤损失。

七、结论

回顾起来，金融危机为发展中国家大城市集聚区社会与经济快速发展所带来的机遇和问题提供了重新思考和定位的机会。全球及地方在经济、政治、社会、文化、空间关系上的变化都对雅加达产生影响，有些人获得成功，有些人则遭遇失败。雅加达未能有效地消除上述变化给城市带来的冲击，城市中出现了社会空间环境的对立与分隔。未来城市需要在发展战略中更多地思考可持续发展的道路，提供更加有效的城市管理与均衡发展手段，减少由于应对全球化竞争和地方多样化要求而产生的社会空间对立。

参考书目

[1] Abayasekere, S. Jakarta a History[M].Oxford：Oxford University Press，1987.

[2] Chen F. L. Emerging World Cities in Pacific Asia [M]. Tokyo：United Nation University Press，1996.

[3] Dick H., Rimmer P. Beyond the Third World City：the New Urban Geography of Southeast Asia [J]. Urban Studies, 1998, 35（12）：2303-2321.

[4] Dinas Tata Kota. RUTR Jakarta 1985-2005 [Z]. Pemerintah Daerah Khusus Ibukota Jakarta，1987.

[5] Dorleans B. Etude Geographique de Trois "Kampung" a Djakarta [Z]. l' Université de Paris-Sorbonne，1976.

[6] Firman T. From "Global City" to "City of Crisis": Jakarta Metropolitan Region Under Economic [Z], 1999.

[7] Turmoil. Habitat International [J]. Elsevier Science, 23 (4): 447-466.

[8] Giebels Lambert J. JABOTABEK, an Indonesian-Dutch Concept on Metropolitan Planning of The Jakarta Region [M]. Dorsrecht-Holland: Foris Publications, 1986.

[9] Hoff V. D. R, Steinberg F. The Integrated Urban Infrastructure Development Program and Urban Management Innovations in Indonesia, I. H. S. Working Paper, Rotterdam [Z], 1993.

[10] Hoffman M. J. Modernizing Land Administration in Indonesia [J]. The Urban Institute for Badan, 1991.

[11] Pertanahan Nasional/ National Land Agency, and U. S. Agency for International Development, Jakarta [Z].

[12] Jellinek L. The Wheel of Fortune: the History of a Poor Community in Jakarta [M]. Allen & Unwin Australia, 1994.

[13] Kusno A. Behind the Postcolonial, Architecture, Urban Space, and Political Cultures in Indonesia [M]. London: Routledge, 2000.

[14] Nas Peter J. M. Issues in Urban Development, Case Studies from Indonesia [M]. CNWS Publication, 1995.

[15] Nas Peter J. M. The Indonesia City [M]. Foris Publication, 1986.

[16] Nas Peter J. M., Veenma M. Toward Sustainable Cities: Urban Community and Environment in the Third World [J]. International Journal of Anthropology, 1998, 13 (2) 101-115.

[17] Nas Peter J.M. Jakarta: Social Cultural Essay [M]. Leiden: KITLV Press, 2000.

[18] Pangestu M. The Boom, Bust, and Restructuring of Indonesian Banks, IMF Working Paper [Z], 2002.

JAKARTA

Andrea Peresthu, Devisari Tunas

For almost four decades, Jakarta was characterized by rapid urbanisation. The rate of urbanisation and its spill over that was steered by free market determinism, has rapidly transformed the city-region into the biggest megapolitan in Southeast Asia. Urbanization grew at the rate of 3.16 percent per annum. The conurbation of Jakarta tripled in three decades, from 4.5 million inhabitants in 1970 to almost 10.5 million in 2006. It holds approximately 32.8 percent of Southeast Asia's urban population, and it is becoming the first large urban agglomeration in the Southeast Asian region. This large period of economic growth went side by side with urbanization and a huge capital concentration in the Jakarta region; it created new spatial settings and a new system of regions, cities and municipalities.

From Dutch till Japanese occupation

During the colonial period, Jakarta was not developed extensively. The amount of infrastructure, housings and public facilities that was built was mainly aimed at supporting the colonial exploitation and production system (i. e. railway from hinterlands to the port and the expansion of the port), instead of improving the living condition of the entire citizenry. Although there was a feeble attempt to make better living conditions of indigenous settlements, such a program were mainly aimed at preventing unhealthy conditions, the city would generate especially social and political tension leading, which could have led to instability.

During the Japanese occupation (World War II), the city was characterized by a huge amount of illegal land occupation, which brought the basis for the current urban development problems. Due to food shortages during the war, the military authority ordered mass food cultivation in all vacant land including public parks, without necessary paying rent or even asking permission from its legal owners. This kind of forced land occupation considerably increased the number of squatter areas, and led to a chaotic land ownership situation until at present.

Nation-Building and Urban Projects

At the beginning of the Soekarno regime, the first president renamed Batavia as Jakarta. This period was characterized by the rise of state-capitalism, where the state managed the public sectors (finance, infrastructure, industry, mining and trading). A huge amount of revenue during this state-capitalism period was aimed at repaying foreign loans, and providing capital for short and long term public investments (i. e. public works, transport, health, education and housing), which was meant to promote the national welfare (1960 National Eight Year Plan). Relaying on the urgency of creating jobs as well as the requirements for new infrastructure to steer economic development, the government erected a number of public works on buildings (e. g. sports and conventional centres, high-rise office buildings) and infrastructures (e. g. The Jakarta-Bogor highway and the north-south ring road). As a result of these developments, the city transformed vastly. New urban forms together with effects followed such a growing number of unplanned urban settlements appeared along several new development corridors (Jakarta-Bogor/Jakarta-Bekasi).

The Imperative of Growth and Its Limitations

The first free market processes during the stabilisation period, fuelled by huge oil revenues (1967-1972) resulted in significant progress in the macroeconomic setting. There was a considerably increase of inflows of foreign investment in manufacturing together with the services and tourism sector, and most of them were taking place in Jakarta. Subsequently, the city needed more labourers. It was manifested by a significant increase in Jakarta's urban population, subsequently, the city's population increased from 2.9 million inhabitants in 1961 to 4.5 million inhabitants in 1971, with an average growth rate of 4.46 percent per annum. The urban core spilled into adjoining areas and was followed by the extension of the administrative boundary from 577km^2 in 1961 to 639km^2 in 1971, an increase of about 10 percent.

However, the rate of urbanization seemed to overtake the limit of urban guidelines that barely existed before, while the city was being confronted with a rapid urban growth. This constraint meant that there were inadequate urban services. Short-term improvements to urban infrastructure were unable to provide better services to overcome the effects of rapid urbanisation. Moreover, the city also had no adequate spatial guidelines that could manage this rapid urban change. The only detailed planning instrument that existed at that time was the Town Planning Act (Wet op de Ruimtelijke Ordening, 1948). This instrument, however, could only deal with a small-scaled town in the rural areas; it was not suitable for guiding an expanding city like Jakarta at that moment.

The Dream of Bundle De-Concentration Metropolis

Under these circumstances, the Ministry of Public-Works together with the Netherlands Directorate for International Technical Assistance attempted to address these rapid development trends by developing detailed guidelines for the Jakarta metropolitan area. The plan aimed at addressing the main factors of urban growth and to consolidate the existing urban structure as well as to anticipate future growth. As a result, in 1973, the first metropolitan growth guideline was introduced, the so-called JABOTABEK, which stands for Jakarta, Bogor, Tangerang and Bekasi. A planning approach addressing the absorption capacity of new settlements within the Jakarta Metropolitan Region.

The JABOTABEK metropolitan plan was obviously inspired by the Dutch metropolitalisation model, the Randstad "concentrated de-concentration". The new plan was based on two basic notions of the growth strategy, namely "bundled de-concentration" and the development of self contained "growth centres". The de-concentration strategy, which contains growth centres with the main emphasis on the agricultural and small-scale manufacturing sectors, aimed at promoting growth of the adjoining depressed region in BOTABEK. Growth centres, based on the planning concept, must be self-contained centres, self-sufficient in terms of employment and services. By implementing this concept, it was expected that the welfare of all income groups in depressed regions would increase and regional disparities would diminish. Social stability would therefore be achieved in the long run. Furthermore, Jakarta as an urban core was supposed to adopt a concentrated strategy that entailed a huge over-reservation of large-scale manufacturing development and services.

Free-Market determinism: the Risk of Urban Society

However, the growth centres planning concept became blurred in spatial-practice, in particular during the free market determinism period. The process led to market liberalisation, decentralisation, deregulation and privatisation (Washington consensus, 1989). This process subsequently stimulated a huge demand for land and property. It facilitated urban expansion to double the area in three decades (from 340km^2 in 1971 to 661km^2 in 1997). This property market boom was to be seen in the considerable changes to road infrastructure system, the consolidation a new CBD, and the overall expansion of the urban fringe consisting of low-density housing, industrial estates and shopping malls. The constraints in land use were to be seen in the rise of spatial crisis, such as the increase of massive land acquisition of low-income dwellers both who reside nearby CDB zones and in the outskirts.

Furthermore, financial deregulation that took place in 1988 enabled the banking sector to expand credit into the property sector. This led to the booming of real estate and considerable growth on the outskirts. Deregulation in the automotive sector reduced import tax, lowered the price of cars, leading to enormous car purchases, which increased mobility and commuter trends. Infrastructure privatisation, such as can be seen in the construction of new highways, also contributed to the pace of urbanization in the outskirts. New highways became well connected both to the urban core and its fringe. Some even reached isolated areas. Hence, the territory of this metropolitan area changed considerably, commuter trends increased as well as traffic jams, socio-spatial segregation also appeared.

In brief, the integration of the local economy to global restructuring has brought about significant wealth accumulation in the urban areas, which was manifested in significant transformations of urban forms of this metropolitan region. However, since major development were mainly concentrating on the real-estate sector, which provided more short-term and quick profits for the private sector, the development of public housing became insignificant. As a result, socio-spatial segregation, which contained a future risk for society increased. It was characterised by many low-cost housings projects being built in some remote terrain, isolated, lack of accessibility and no public transport, vulnerable to floods, poor infrastructure as well as low-standard living conditions.

The Price of Global Integration

By mid 1997 the Asian financial crisis hit Indonesia badly, and the burgeoning segregated city witnessed an unprecedented multi dimensional crisis. The economic crash led to project rescheduling and subsequently led to discontinued capital inflows to almost all construction projects. Within months, many developers got bankrupt, the few that survived rescheduled, postponed or even cancel their projects. As a result of project cancellation, unemployment escalated and has become one of the trigger factors in the complex socio-political crisis in Jakarta. The worsening economic situation and displeasure with severe short-term social and political reforms have in turn prompted some of the biggest urban riots in Jakarta. Consequently, for a short while, the city's economy suffered. Thus the city lost its competitiveness. Considerable capital flights occurred and prompted multinational manufactures to relocate to other more accommodating countries in the region.

Conclusion

Recent economic recovery seems to have achieved moderate growth. However, global conditions related with local socio-political settings, such as growing terrorism, radicalisms, labour movements and separatisms have contributed to serious uncertainties which in turn has reduced the competitiveness of Indonesia compared to other countries in the region, such as Vietnam and China. Even though, recently, government has offered various incentives, new infrastructure development (i. e. Large Urban Projects) to encourage capital inflows; many multinational companies seems less interested in investing or even consolidating their centre of production from Indonesia, due to a prolonged socio-political crisis that could worsened the situation.

In brief, this rapid growth does not imply that spatial transformations conform to spatial guidelines toward sustainable paths. Violations of the master plan have become increasingly uncontrollable, as it receives support and privileges from local elites. The impact is clearly visible and has resulted in a disproportioned urbanisation process. To put it bluntly, the "achievement" of this rapid growth does not comply with objectives to promote sustainable and even growth, as is prescribed by the growth model of the Jakarta metropolitan region. Furthermore, it even stretches the socio-spatial segregation model.

It is therefore necessary to seek an appropriate development model that could on one hand manage the shortage of current urban problems (i. e. affordable housing and sustainable public transports system), and on the other, build accommodating spatial guidelines and increasing law enforcement due to avoid violation of these guidelines.

The imperative of growth should be confronted with scientific reflections and questions on how to manage the huge urban scale in a more sustainable way as well as to foresee future development implications that would affect this metropolitan area. It is, therefore, imperative to seek a new political agreement and platform for guiding the process of future developments that intends to simultaneously improve the living conditions of all citizens and upon the awareness that this region is always being confronted with limited resources (in terms of capital, human, infrastructures, etc.).